Deutsches Zentrum für Entwicklungstechnologien – GATE

Deutsches Zentrum für Entwicklungstechnologien – GATE – stands for German Appropriate Technology Exchange. It was founded in 1978 as a special division of the Deutsche Gesellschaft für Technische Zusammenarbeit (GTZ) GmbH. GATE is a centre for the dissemination and promotion of appropriate technologies for developing countries. GATE defines „Appropriate technologies" as those which are suitable and acceptable in the light of economic, social and cultural criteria. They should contribute to socio-economic development whilst ensuring optimal utilization of resources and minimal detriment to the environment. Depending on the case at hand a traditional, intermediate or highly-developed can be the „appropriate" one. GATE focusses its work on four key areas:

– *Technology Exchange*: Collecting, processing and disseminating information on technologies appropriate to the needs of the developing countries; ascertaining the technological requirements of Third World countries; support in the form of personnel, material and equipment to promote the development and adaptation of technologies for developing countries.

– *Research and Development:* Conducting and/or promoting research and development work in appropriate technologies.

– *Cooperation in Technological Development:* Cooperation in the form of joint projects with relevant institutions in developing countries and in the Federal Republic of Germany.

– *Environmental Protection:* The growing importance of ecology and environmental protection require better coordination and harmonization of projects. In order to tackle these tasks more effectively, a coordination center was set up within GATE in 1985.

GATE has entered into cooperation agreements with a number of technology centres in Third World countries.

GATE offers a free information service on appropriate technologies for all public and private development institutions in developing countries, dealing with the development, adaptation, introduction and application of technologies.

Deutsche Gesellschaft für Technische Zusammenarbeit (GTZ) GmbH

The government-owned GTZ operates in the field of Technical Cooperation. 2200 German experts are working together with partners from about 100 countries of Africa, Asia and Latin America in projects covering practically every sector of agriculture, forestry, economic development, social services and institutional and material infrastructure. – The GTZ is commissioned to do this work both by the Government of the Federal Republic of Germany and by other government or semi-government authorities.

The GTZ activities encompass:

– appraisal, technical planning, control and supervision of technical cooperation projects commissioned by the Government of the Federal Republic or by other authorities

– providing an advisory service to other agencies also working on development projects

– the recruitment, selection, briefing, assignment, administration of expert personnel and their welfare and technical backstopping during their period of assignment

– provision of materials and equipment for projects, planning work, selection, purchasing and shipment to the developing countries

– management of all financial obligations to the partner-country.

Deutsches Zentrum für Entwicklungstechnologien – GATE
in: Deutsche Gesellschaft für Technische Zusammenarbeit (GTZ) GmbH
Postbox 5180
D-6236 Eschborn 1
Federal Republic of Germany
Tel.: (06196) 79-0 Telex: 41523-0 gtz d

Klaus Kuhnke, Marianne Reuber, Detlef Schwefel

Solar Cookers in the Third World

Evaluation of the Prerequisites, Prospects and Impacts of an Innovative Technology

Springer Fachmedien Wiesbaden GmbH

The authors:
Dr. Klaus Kuhnke is professor of applied physics at the Fachhochschule of Osnabrück, Germany. He taught solar energy at the Engineers' School of Tunis and worked for some years in industry and research institutions in the field of thermal solar energy applications.
Marianne Reuber is a graduate in education science, majoring in education in the Third World and adult education. She has worked for the GTZ since 1986 as an appraiser.
Prof. Dr. rer. pol. Detlef Schwefel works for over twenty years on socio-economic aspects of health and health care. His work comprised planning and evaluation of health and other projects in developed and developing countries, among them work on health indicators, evaluation methods and socio-economic risk factors.

Second, revised and translated edition, 1990

Deutsche Bibliothek Cataloguing-in-Publication Data

Kuhnke, Klaus:
Solar cookers in the third world : evaluation of the prerequisites, prospects and impacts of an innovative technology ; a publication of Deutsches Zentrum für Entwicklungstechnologien (GATE) in Deutsche Gesellschaft für Technische Zusammenarbeit (GTZ) GmbH / Klaus Kuhnke ; Marianne Reuber ; Detlef Schwefel. – 1. Aufl.
ISBN 978-3-528-02056-9 ISBN 978-3-663-13939-3 (eBook)
DOI 10.1007/978-3-663-13939-3
NE: Reuber, Marianne:; Schwefel, Detlef:

The author's opinion does not necessarily represent the view of the publisher.

Originally published by Friedr. Vieweg & Sohn Verlagsgesellschaft mbH, Braunschweig in 1997

ISBN 978-3-528-02056-9

Table of Contents

Key terms: Solar cooker: technology, diffusion, evaluation, acceptance, projects

List of Photos/Illustrations

List of Tables

Types of solar cookers

Box-type solar cookers:

ATDO reflector box type solar cooker
ATRC solar cooker
Cocina solar
Dhauladhar solar cooker
Indian box type solar cooker
Kerr-cole solar box cooker
MECTAT solar ovens
Mina solar pressure cooker
Orangi cooking box
RERI-SEP solar cooking box
RIIC solar oven
SERVE solar oven
Suryamuklu box type solar cooker
22° solar cooker
ULOG tropical solar cooker
ULOG European solar cooker
Four-mirror cooking box

Reflector cookers:

Advanced reflector cooker for Mali
Bottom-heated concentrator box
External concentrating eccentric axis box style solar cooker (EEB)
Falco S/C
Sobako 1 / SOBA 1
Sun basket
Sungril
Suryakund
Table-type reflector cooker
Tube solar oven
Valparaiso reflector cooker
VIAX solar cooker SK 10

Heat accumulating solar cookers:

Heat accumulating steam cooker
ISE solar cooker with integral oil storage
Solar hot plate cooker
Heat pipe storage solar cooker with evacuated tube collectors

Convective cookers:

Convective solar cooker (CSC)
Steam immersion heater solar cooker

Comparative survey:

Synoptic comparison table of solar cookers

Symbols and Units

A	aperture, area of incidence	m^2
a	air permeability coefficient	$m^3/(h\ m\ Pa^{2/3})$
b	breadth, width	m
c	specific heat capacity	kJ/(kg K)
CC	cost of cooking, price of energy per kg of food cooked during the first year	$/kg
CCL	cost of cooking with respect to the lifetime, price per kg of food cooked over the lifetime of the cooker	$/kg
d	depth	m
G	global irradiation	MJ/m^2 ; kWh/m^2
G', g	global irradiance	W/m^2
H	diffuse (sky) radiation	MJ/m^2 ; kWh/m^2
H'	specific enthalpy	kJ/kg
ΔH'	differential specific enthalpy	kJ/kg
h	height	m
L	useful lifetime in years	a
l	length	m
l_i	length of joint i	m
m	mass	kg
n	number of meals cooked per day	1
P	purchase price	$
PEL	lifetime price of energy	$/kWh
p	pressure	Pa
Δp_i	pressure drop across joint i	Pa
Q'	thermal output power	W
S	direct (beam) radiation	MJ/m^2 ; kWh/m^2
T	temperature	°C
ΔT	temperature difference	K
ΔT'	temperature interval of the cooking process	K
t	time, cooking time	s, h, d, a
U	heat transfer coefficient	$W/(m^2K)$
V	volume	m^3; l
V'	volumetric flow	m^3/s
α	absorptance	1
η	efficiency	1
η'	efficiency of energy-collecting components	1
λ	wavelength	μm
ρ	density	kg/m^3
τ	transmittance	1

Indices

A	aperture
i	counting index
J	joints
L	loss
Opt	optical
U	useful
W	wall

1. Introduction

Hopes: Some people see cooking with solar energy as a neat way to alleviate or even solve several development problems at once:

- women's exposure to unhealthful cooking conditions, and their hard, time-consuming work of collecting fuel
- shortage of firewood (read: the energy crisis of poor people in the Third World) with such devastating ecological consequences as deforestation, rampant soil erosion and subsequent desertification
- dependence on nonrenewable sources of energy, with attendant balance-of-payment problems for the nation as a whole
- a lack of future-oriented technologies that would be fit for small-scale application and therefore (could) have a gigantic global market.

Solar cooking also touches upon a basic need: nutrition. At first glance, cooking with solar energy has the appearance of a development-political stroke of luck.

Disappointments: While various solar cookers have been developed and tested, practically all attempts to get appreciable numbers of them put into actual service have been unsuccessful. Are solar cookers still technically immature, or do they constitute inappropriate technology? Why are they (still) not being accepted by people in developing countries?

Background: This report attempts to answer such questions. In addition to the results of an intensive study of pertinent literature, the report is based mainly on a cross-section analysis conducted on behalf of Deutsche Gesellschaft für Technische Zusammenarbeit (GTZ) GmbH* and the German

* German Agency for Technical Cooperation

Appropriate Technology Exchange (GATE). The analysis, conducted by the authors themselves in 1986/87, included taking stock - by way of a worldwide postal survey - of all solar cookers developed by 1986/87, in addition to diagnosing solar cooker projects in India, Kenya, Mali, Pakistan and the Sudan. The cross-section analysis attaches special importance to the socioeconomic, psychosocial and sociocultural conditions of acceptance, although the technical aspects are also given due attention, of course.

Intentions: This report addresses development-political decision-makers as well as engineers and technicians engaged in innovative work aimed at harnessing solar energy. Its main objective is to draw attention to the many socioeconomic, sociocultural and psychosocial technical framework conditions that can be decisive for the dissemination of new ideas and/or the diffusion of innovations. In that sense, this report assesses the secondary effects of a soft technology and analyses its prerequisites, prospects and risks. The following interdependences are of crucial importance:

- Poverty: "A central function of development aid, one which in the past has been inadequately realized, is to reach the poorest social strata" /24,7/**. "Development aid provided by the Federal Republic of Germany is directed primarily at the poor social strata" /24,20/. Thus read the Federal German Government's currently valid guidelines on development policy. Accordingly, any solar cooker project must stand judgment with regard to how much it can contribute toward those ends.

- Sources of energy: Substituting indigenous energy for imported energy can be very helpful for improving a country's balance of payments by lessening its dependences, reducing its foreign-exchange shortage and easing its foreign-debt burden. This is all very important for the country's overall

** The first number in brackets refers to the position in the list of references, and the number behind the comma indicates the page.

economy. But when poor people are the subject of discussion, the question of household energy supplies is the crucial point ... "Cooperation in the field of new energy technologies within the energy sector in general as a means of expanding the technological capabilities of developing countries" /24,29/, which the Government of the Federal Republic of Germany both calls for and promotes, must - in order to be in line with that country's prime development-political objective - place first emphasis on decentralized forms of energy supply, "which stands to improve the living conditions of large sections of the population, first and foremost the rural population, while preserving natural resources, particularly forests and soil fertility. A broad effect should be achieved by offering uncomplicated equipment and appliances that can be kept in good working order by the native users and which prove to be cost-efficient at their respective places of use. To that end, the aid of suitable national-scale technoscientific and/or application-oriented sponsors is enlisted for improving basically economical technologies, field testing of technically mature technologies for inservice economy, and ascertaining the actual degree of sociocultural acceptance" /22,22/. Thus stipulates the Federal Government's Program for Cooperation with Developing Countries in the field of energy technology.

- <u>Diet</u>: Some important forms of energy are often overlooked: manpower, for instance, i.e. the energy provided by human muscles and the energy content of the nourishment needed to replenish the strength of those muscles. Most people's thoughts tend to center on technical sources of energy. But when energy per se and its rational use are to be promoted via development-political considerations, then the links between primary, secondary and tertiary energies and between technically relevant and humanly relevant energies are necessarily of primary interest. In a joint statement, the Federal Ministry for Economic Cooperation (BMZ) and the Federal Ministry of Food, Agriculture and Forestry announced that the Federal Government attaches very special importance to "making a lasting, effective contribution toward improving the global nutritional situation and toward socioeconomic development in the Third World" /20,III). Solar cooker projects must be seen in that light and judged accordingly.

- Women: Practically everywhere, women, not men, do the cooking. Solar cooker projects therefore address women directly. Consequently, a general requirement for all measures is that they "contribute directly toward improving the situation of women in their role as housewife/mother and producer/employee" and toward "advancing their social self-determination and political participation" /43,23/, most notably through
  - lightening their household workload
  - improving their housekeeping methods
  - family health and childcare
  - promoting the economic/financial activities of women /43,23ff/.

  Solar cooker projects therefore should display a high degree of relevance to women. Whether or not that is the case, and the extent to which solar cookers might even aggravate women's work, are issues of special importance.

- Technologies: Solar cookers are an example of technological innovation. In view of widespread negative experience in the past, appropriate technology is in increasingly high demand for the developing countries. "Robustness, durability and easy maintenance" /49,33/ are prime criteria. Research and development in this area must include comprehensive field tests to determine how well or poorly a given technology fits into its overall system context: environment, engineering, economy, social aspects, politics, culture, values; assessing the impact of technology is a basic element and concomitant aspect of developing appropriate technologies, as called for by GTZ /49,38/. For solar cookers, too, the main question is: Is the technology appropriate, i.e. does it fit the needs and living conditions of the target groups of development policy? Thus, according to the development-political guidelines followed by the government of the Federal Republic of Germany, "development policy must have as close a tie-in as possible to the given cultural and social situation" /24,12/.

<u>Systems</u>: Technological innovation needs integration into political, socioeconomic, cultural and ecological networks. To limit one's investigation to certain predetermined parts of that network would be scientifically unjustifiable. In other words, any evaluation must aim first for breadth - and even sacrifice depth if need be - in order to gain relevance. Then, in its second phase, it may begin to address other, rationally chosen matters. Since evaluative research on solar cookers is just getting started, its systematic, exemplary expansion in the sense of a technology impact assessment would be a worthwhile task for future development-political research.

2. Solar Cookers and Solar Cooker Projects

This chapter characterizes the four basic types of solar cookers, describes the physical fundamentals of solar cooking, and scrutinizes solar cooker projects in India, Kenya, Mali, Pakistan and the Sudan. Subsequently, it examines the present state of the art and dissemination. The emphasis is on technical aspects.

2.1 Types and Techniques

Solar cookers and ovens, absorb solar energy and convert it to heat, which is then used for cooking or baking various kinds of food.

Differentiation is made between four basic types of solar cookers:
- box-type solar cookers
- concentrating-type or reflector cookers
- heat-accumulating solar cookers
- solar steam cookers.

There are also some hybrid types, e.g. heat-accumulating solar cooking boxes with reflectors. Various types of solar cookers are described in the appendices.

2.1.1 Box-type solar cookers

Box-type solar cookers - or more precisely: solar cooker-cum-ovens (or solar ranges) - consist of a well-insulated box with a black interior, into which black pots containing food are placed. The cover of the box usually comprises a two-pane "window" that lets solar radiation enter the box but keeps the heat from escaping. This in addition to a lid with a mirror on the inside that can be adjusted to intensify the incident radiation when it is open and improve the box's insulation when it is closed.

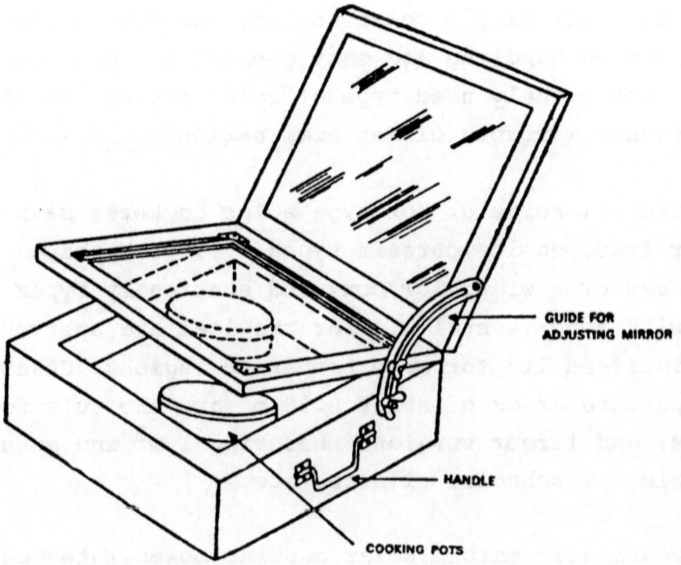

Figure 1: Box-type solar cooker

The main advantages of box-type solar cookers are:

- They make use of both direct and diffuse solar radiation
- Several vessels can be heated at once
- They can double as an oven (not for crispy baked goods)
- They are light and portable
- They are easy to handle and operate
- They needn't track the sun
- The moderate temperatures make stirring unnecessary
- The food can be kept warm until evening
- The boxes are easy to make and repair using locally or regionally available materials
- They are relatively inexpensive (compared to other types of solar cookers).

There are some disadvantages too, of course:

- Cooking must be limited to the daylight hours
- The moderate temperatures make for long cooking times
- The food is not accessible for stirring, turning, etc.
- The glass cover causes considerable heat losses
- Such cookers cannot be used for frying or grilling.

Thanks to their simple construction, relatively low cost, uncomplicated handling and easy operation, solar cooking boxes are the most widely used type of solar cooker. Their practical significance warrants closer examination.

There are all sorts of box-type solar cookers: mass-produced, hand-crafted, do-it-yourself types ..., with shapes resembling a suitcase or a wide, low box, and stationary types made of clay, with a horizontal lid for tropical and subtropical areas or an inclined lid for more temperate regions. Standard models with aperture areas of about 0.25 m^2 are the rule for a family of five, and larger versions measuring 1 m^2 and more are available for schools, clinics, etc.

Many technically mature solar cooking boxes intended for use in the tropics (high solar altitude) are of shallow design. The inner box, usually made of blackened sheet metal, must be watertight - or better: vaporproof - with respect to the insulation. The bottom must also be painted (or otherwise coated) flat black to achieve good absorption; the sides of some types (Indian, for instance) are also black, while other designs prefer reflecting walls (ULOG cookers).

Since the heat absorbed by the inner box needs to be conducted to the area beneath the cooking pots, the best choice of material is aluminum, because it is a very good heat conductor. Additionally, aluminum is good for reasons of corrosion prevention, i.e. iron sheet boxes, even galvanized ones, could not stand up indefinitely to the hot, humid conditions that are created inside during the cooking process. Sheet copper is prohibitively expensive. Oehler uses old aluminum offset-printing plates as an inexpensive source of material for the inner boxes of his ULOG solar cooker/ovens /130/.

No metal parts should extend to the outside around the top rim of the inner box: thermal bridges must be avoided. The insulation may consist of spun glass, rock wool or some natural material like residue from the processing of peanuts, coconuts, rice, corn, etc. Whatever kind of material is used, it must be kept dry.

The cover must consist of two panes of glass with a layer of air between them. The pane-to-pane clearance usually amounts to 10...20 mm. Recent experiments have shown that a honeycomb structure of transparent material that divides the interspace into small vertical compartments can substantially reduce the cooker's heat losses, thus increasing its efficiency accordingly /145/. The inside cover pane is exposed to substantial amounts of thermal stress, for which reason tempered (safety) glass is frequently used; otherwise, both panes may consist of normal window glass with a thickness of about 3 mm.

Set in a (wooden) frame, the two panes of glass yield a tiltable cover. The glass must be held flexibly and free of stress in the frame; the joint must be heat-resistant up to 150°C (e.g. silicone). The double-pane cover must fit tightly around the top of the box in order to trap as much heat as possible, thus maximizing the cooker's efficiency and minimizing the required cooking time.

The outer cover, or lid, of the solar cooking box always serves as a reflector to amplify the incident radiation. The reflecting surface may consist of an ordinary glass mirror (heavy, expensive, fragile, but easily obtainable anywhere), plastic sheet with a reflecting coating (Mylar, Tedlar, etc.; cheap, but not very durable and hard to find), or a metal mirror (unbreakable, but dulls easily). In an emergency, even foil from empty cigarette packs will do the job.

A simple catch device keeps the lid in the proper reflection position (depending on the altitude of the sun), and a latch is used to close it like a suitcase. Both the positioning mechanism and the latch must be fail-safe to the extent that they cannot get between the lid and the cover and break the glass. The closed lid - with or without a layer of insulation behind the mirror - keeps the food warm. The hinges for the reflecting lid must be strong enough and carefully mounted to ensure that they stand up to everyday use: the hinges are a frequent weakpoint. Instead of metal hinges, strips of leather or textile can be used (cf. Appendix 2, item no.13).

The outer box of the solar cooker may be made of wood, glass-reinforced plastic (GRP) or metal. GRP is light, inexpensive and fairly weather-resistant, but not necessarily stable enough for continuous use. Wood is more stable, but also heavier and less weather-resistant. A metal case (aluminum) with wooden bracing offers the best finish (the cooker's image and appeal are important acceptance factors) and is adequately stable with regard to mechanical impact and the effects of weather. An aluminum-clad wooden box is the most stable of all, but it is expensive and time-consuming to make, in addition to being heavy.

All box-type solar cookers have one or more handles on the sides or front; some have castor wheels. Still others are of the stationary type, i.e. home-built clay structures that are extremely inexpensive (Dhauladhar solar cooker), but which have the disadvantage of not being able to track the sun during the morning and afternoon hours. In addition, special precautionary measures must be taken to get the cooker through the rainy season unscathed.

The basic shape of all solar cooking boxes is rectangular (with the exception of the Franco-Indian one-pot rice cooker, which is round). A quadratic design minimizes heat losses through the joints and side walls, while a rectangular design has other advantages: the hinges are exposed to less wear & tear and the long reflecting lid makes solar positioning less critical (cf. ATRC solar cooker in Appendix 1).

Suitable cooking pots are important prerequisites for successful solar cooking. Shallow flat-black aluminum vessels with flat bottoms that make good thermal contact with the metal floor of the inner box are most suitable. The pots used in India are no more than 7 cm high and have flat, tightly closing lids. According to Indian experience, the bottom pane of glass in the cover should be about 2 cm above the lid of the pot. A smaller clearance would tend to aggravate losses by heat conduction, and a larger clearance would promote convection with largely the same results.

According to Parikh /136/, there are three ways to effect heat transport to the food. These are:

1. absorption by the sheet-metal bottom, heat conduction into the area below the pot, heating of food from below
2. absorption of solar radiation by the lid and other exposed parts of the pot
3. heating of the food by convection of the hot air surrounding the pot.

Oehler reports that a grid supporting the pot 1...2 cm above the box bottom, increases convective heat transfer considerably and improves the efficiency of the cooking box. If the cooking vessel contains only a shallow layer of food, heating occurs quite rapidly, and the food requires less cooking time. The optimal pot filling heigt is 2...3 cm, or roughly 1 inch.

The capacity of a normal box-type solar cooker with a 0.25 m^2 area of incidence (aperture) amounts to 2 kg solids, i.e. about 4 kg ready-to-eat food, or enough to feed a family of five.

The inside of a solar cooking box can reach a peak temperature of over 150°C on a sunny day in the tropics; that amounts to a thermal head of 120 K, referred to the ambient temperature. Since the water content of food does not heat up beyond 100°C, a loaded solar cooker will always show an accordingly lower inside temperature.

Figure 2 shows a typical time history of temperature in a food-containing solar cooking box /138/.

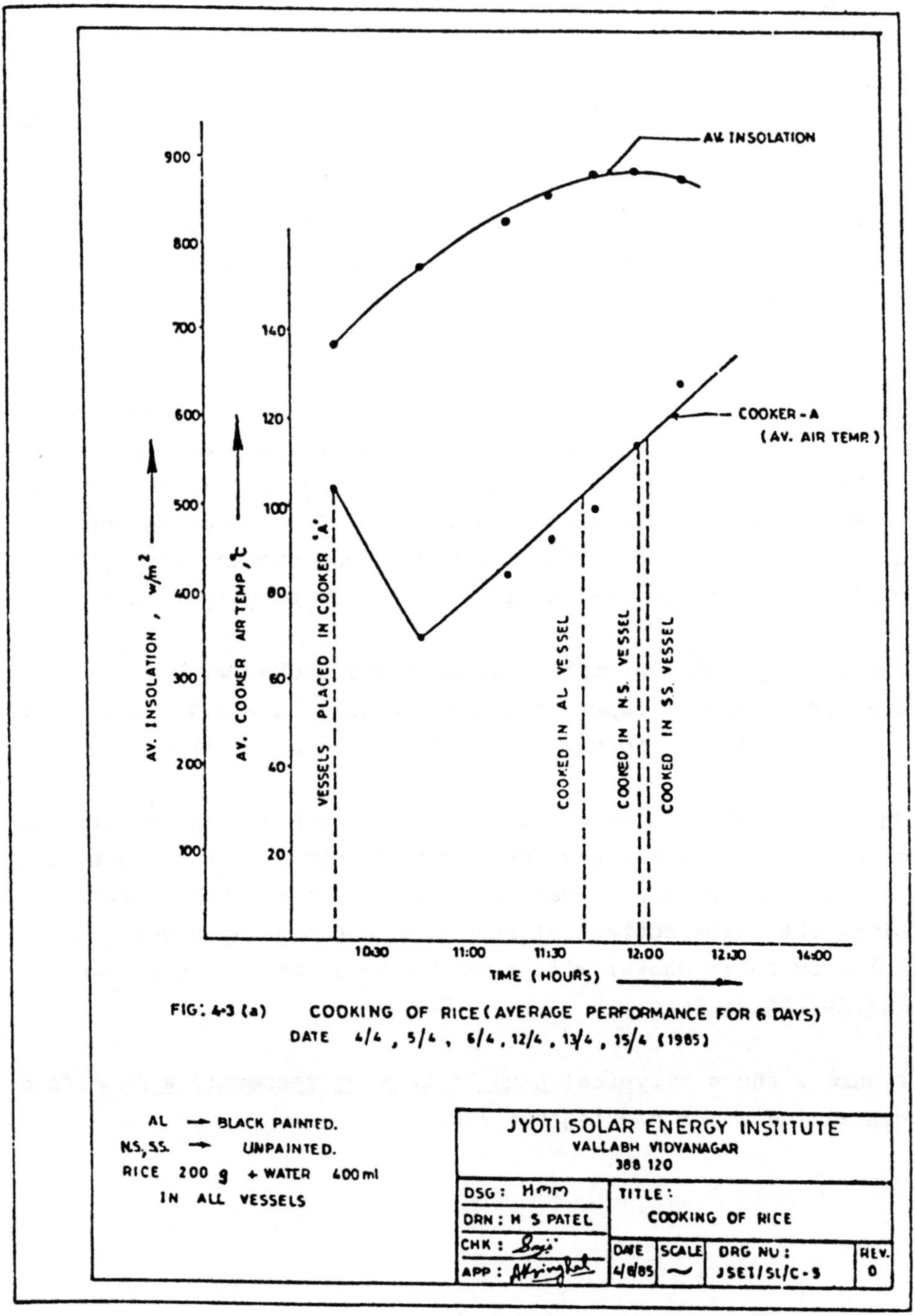

Figure 2: Global radiation and air temperature in a food-containing solar cooking box vs. time

The temperature inside of the solar cooker drops off sharply when the vessels are placed inside it. Also conspicuous is the fact that the temperature remains well below 100°C for the greater part of the cooking time. The boiling temperature of 100°C is not necessary for most vegetables and cereals.

The average achievable cooking times in box-type solar cookers amount to somewhere between 1 ½ and 3 hours for good insolation and a reasonable fill volume. Thin-walled aluminum vessels yield much shorter cooking times than, say, stainless steel pots. The time taken for cooking is also influenced by the following factors, of course:

- The cooking time is shortened by strong insolation and vice versa
- High ambient temperatures shorten the cooking time, and vice versa
- Small volumes (shallow fill) in the pot make for shorter cooking times, and vice versa.

Parikh drew up a cooking factor scheme that combines various factors with an influence on the qualitative and quantitative performance of box-type solar cookers and provides information on the various cooking options and times /75/. The empirical values refer to the western part of Central India (cf. table 1). The basic criteria in Parikh's scheme are

- the weather (season)
- time of day (when cooking begins)
- kind of food
- fill height of food in the pot.

Each of the four main criteria can assume any of three different values, as presented by the multipliers 1, 2 and 3. Once the multipliers have been defined for all four basic factors, the results are multiplied with each other to arrive at a total value situated somewhere between 1 and 81. The results break down into categories that correspond to different cooking times (cf. table 1). Example: vegetables (multiplier 1), 4 cm high in the cooking pot (multiplier 2), placed in the cooking box at about 1:30 p.m. (multiplier 3), requiring a cooking time of 3.5...4 hours in Central India in May (multiplier 1), because the product of multipliers is 2x1x3x1 = 6, meaning 3.5 - 4 hours as per table 1.

Table 1: R.Parikh's cooking-factor/cooking-time multiplication table /75/

Sl. No.	Factors	1 (Multiplier 1)	2 (Multiplier 2)	3 (Multiplier 3)
		Summer	*Moderate*	*Winter*
1.	Weather	April, May, June	October, November, March	December, January, February
		Noon	*Morning*	*Afternoon*
2.	Time of placing food in the cooker	11:00 a.m. to 3.00 p.m.	Upto 3 hours after sunrise 6:00 a.m. to 10:00 a.m.	3:00 to 4:00 p.m.
3.	Kind of food (from cooking point of view)	Soft (vegetables)	Medium (rice)	Hard (dal, meat, etc.)
4.	Thickness of the food material	2 cm ($\frac{3}{4}$ inch)	4 cm ($1\frac{1}{2}$ inch)	6 cm ($2\frac{1}{4}$ inch)

Sl. No.	Product of multiplication factor	Time taken for cooking		
		Hours		Minutes
1.	1	1	—	00
2.	2	1	—	30
3.	3-4	2	—	00
4.	6-8-9	3	—	00
5.	12-16-18-24	4	—	00
6.	27-36-54-81	Food cannot be cooked		

2.1.2 Reflector cookers and concentrators

The most elementary kind of reflector cooker is one that consists of (more or less) parabolic reflectors and a holder for the cooking pot situated at the cooker's focal spot. If the cooker is properly aligned with the sun, the solar energy bounces off of the reflectors such that it all meets at the focal spot, thus heating the pot.

The <u>reflector</u> can be a rigid axial paraboloid, made for example from sheet metal or from a reflecting foil by application of high or low pressure, possibly with some manner of sectionalization provided for taking it apart, folding it together or otherwise rendering it portable. The reflecting surface is usually made of treated aluminum or a mirror-finish metal or plastic sheet, but it may also consist of numerous little flat mirrors cemented onto the inside of the paraboloid. Depending on the desired focal length, the reflector may have the shape of a deep bowl that completely "swallows" the pot (short focal length, pot shielded from the wind) or that of a shallow plate with the cooking pot mounted in the focal point a certain distance above or in front of it.

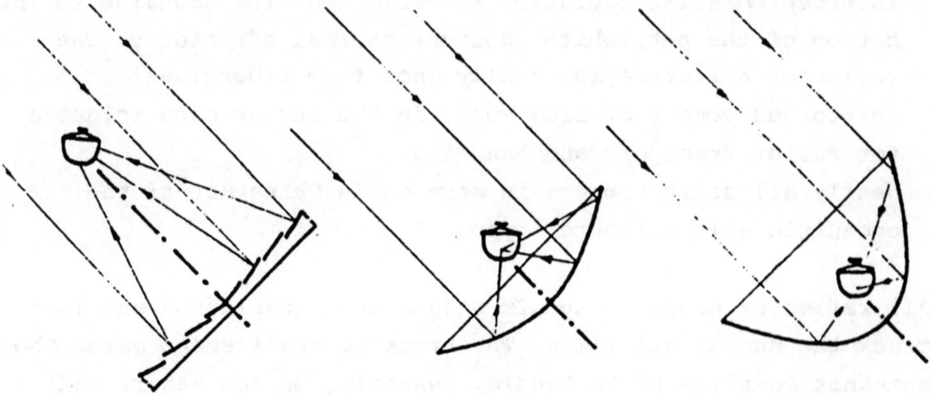

<u>Figure 3:</u> Fresnel-type and other reflector cookers with various focal lengths /126/

A somewhat more detailed introduction to the mechanics of reflector cookers is offered in /5/. Some special designs are touched upon below.

- The <u>Fresnel cooker</u> has a parabolic reflector consisting of several concentric rings arranged in a single plane. This approach facilitates manufacturing and gives the device a lighter, more slender appearance, e.g. the VITA cooker;

- Linear paraboloids, e.g. the Sobako, have a focal line instead of a focal point. Several cooking vessels can be strung out along the focal line;

- Fixed-focus solar cookers are designed such that, as the day progresses, the reflector can be rotated about its own polar axis (parallel to the earth's axis) while the focal point does not change its location at all. Consequently, the cooking pot can be rigidly mounted for added stability and even provided with thermal insulation. The cook can stand in the shade and, with some effort, the "hearth" can even be installed inside the house /95;19/.

- Eccentric axis reflector cookers have reflectors that represent only the lower part of a parabolic dish. Thus, all intercepted solar radiation is reflected from downside to the bottom of the pot, which improves thermal efficiency. The reflector structure is usually made form fiberglass reinforced cement or from wood, in the latter case foldable for easier transport and handling.
  Nearly all solar cookers in service in China are of the eccentric axis reflector type.

All reflector cookers exploit only direct insolation and must track the sun at all times. The tracking requirement makes them somewhat complicated to handle, depending on the nature and stability of the stand and adjusting mechanism.

The advantages of reflector cookers include:
- the ability to achieve high temperatures
- and accordingly short cooking times
- relatively inexpensive versions are possible
- some of them can also be used for baking.

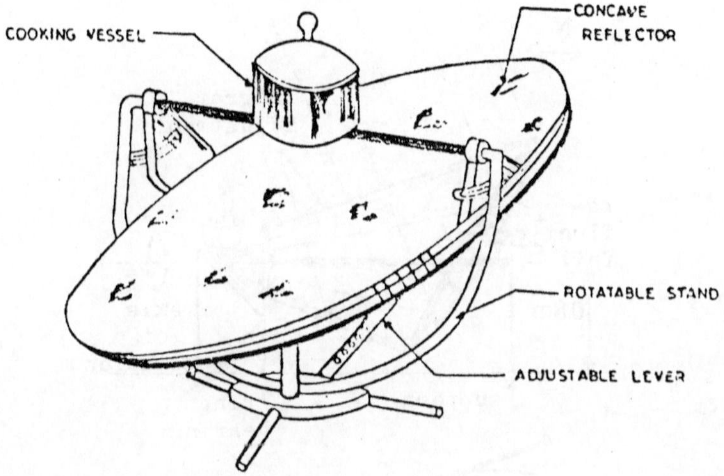

Figure 4: Reflector cooker with pot stand

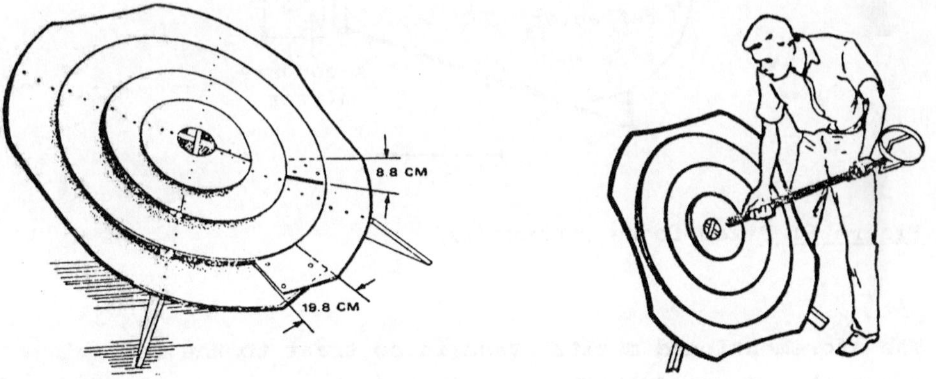

Figure 5: Fresnel-type reflector cooker (VITA cooker)

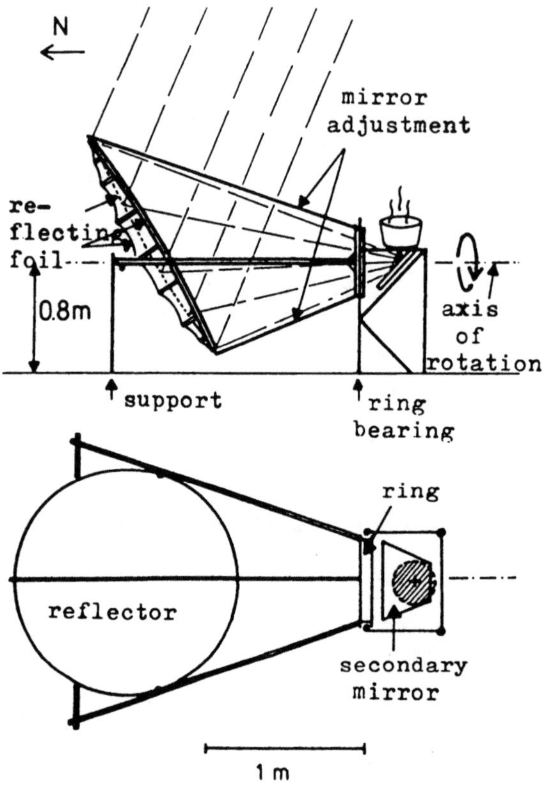

<u>Figure 6:</u> Fixed-focus cooker /19/

The aforementioned merits stand in contrast to the following <u>drawbacks</u>, some of which are quite serious:

- Depending on its focal length, the cooker must be realigned with the sun every 15 minutes or so
- Only direct insolation is exploited, i.e. diffuse radiation goes unused
- Even scattered clouds can cause high heat losses
- The handling and operation of such cookers is not easy; it requires practice, a good grasp of the working principle, and constant close attention to the job at hand
- The reflected radiation is blinding, and there is danger of injury by burning when manipulating the pot in the cooker's focal spot
- Cooking is restricted to the daylight hours

- The cook must stand out in the hot sun (single exception: fixed-focus cookers)
- The reflector is somewhat fragile and its mirror finish normally requires the use of nonlocal materials
- The efficiency is heavily dependent on the momentary wind conditions
- Any food cooked around noon or in the afternoon gets cold by evening.

Particularly the cooker's complicated handling, in combination with the fact that the cook has to stand out in the sun, is a major impediment with regard to the acceptance of reflector cookers. But in China, where the food demands high cooking power and temperature, eccentric axis reflector cookers have been disseminated and accepted in a large number.

Box-type solar cookers, too, can be fitted with reflectors to amplify the incident radiation (booster principle). In most cases, the reflectors are arranged in the form of quadratic or octagonal reflecting funnels that fit onto the top of the cooking box such as to enlarge the effective aperture by a factor of 2...3, resulting in a corresponding increase in energy capture without the need to concentrate the radiation at a focal point. There are booster cooking boxes with and without heat stores.

2.1.3 Heat-accumulating solar cookers

Heat-accumulating solar cookers gather the heat of the sun all day long and store it for use sometime after sundown - or even the next morning. They eliminate one of the main drawbacks of all other solar cookers, which are only useful during sunshine hours. Heat storage, by contrast, extends the cooking option to the time of day when most cooking is done in Third World cultures: in the morning and in the evening.

The simplest type of heat-accumulating solar cooker is an ordinary box-type solar cooker containing a few bricks. That alone is enough to shift the cooking time from the late afternoon towards the early evening. However, the method is not very efficient and really only useful for keeping the food warm somewhat longer.

Strictly speaking, turning a solar cooker into a heat accumulator requires some very special design considerations. Such devices contain substantial amounts of heat-retaining material and are usually designed to withstand very high temperatures, since the mass required to store a certain amount of thermal energy is inversely proportional to the temperature. The weight of the heat store, the heat-transfer mechanism and the (highly efficient or concentrating-type) high-temperature collector make the cooker quite heavy and voluminous, complicated to build, and usually quite expensive.

The more familiar designs use heat stores consisting of iron, magnesite, water or a high-boiling fluid (thermal oil). The Bomin Solar Hot Plate Cooker uses a funnel-shaped reflector to amplify the incident radiation by a factor of roughly 3 in order to heat a set of iron plates to very high temperatures in a sort of cooking box. The iron plates are then removed and transferred to a separate cooking area, where they impart their heat to cooking pots.
The heat-accumulating solar cooker designed by Pöhlmann/Stoy uses high-efficiency evacuated collector pipes to trap solar energy. The heat is transferred by heat pipe to the solid heat store, which then delivers controlled amounts of heat to a hot plate.

The ISE heat-accumulating solar cooker uses a highly efficient oil-filled flat plate collector that operates on the thermosiphon circulation principle to carry heat to an elevated hot-oil storage tank. Gravity circulation transfers the heat to the oil tank without need of a pump. The cooking pot stands in a matching depression in the top of the oil tank. The heat-accumulating steam cooker devised by Mills and Qiu (solar cooking stove) stores heat in pressurized water at more than 100°C. Heat extraction via the cooking pot lowers the system's internal pressure, thus causing the water to boil and carry more heat to the pot according to the heat-pipe principle.

The possibility of storing latent heat in solar cookers has been investigated in detail /99/. Latent-heat stores can collect heat without experiencing an increase in temperature. Instead, they undergo a physical change of state. In principle, this makes it possible to store more heat at a lower temperature and/or in less storage mass than in the case of normal (sensible) heat storage. The lower temperature reduces heat losses, a fact which is of advantage for collecting heat as well as for storing it.

A great number of latent-store substances and mixtures were investigated in /99/. The main preconditions were that the conversion (i.e. latent-storage) temperature had to be situated within a range regarded as acceptable for solar cooking (150...300°C), and the material had to be relatively nontoxic, inexpensive, stable and available. Of all the tested metals, alloys, organic compounds and inorganic saline mixtures, the mixture $NaNO_2/NaNO_3$ emerged as the most favorable (in relation to the others), but there is still a long way to go before a really serviceable latent-heat-storage solar cooker can hit the market. Among the problems still to be remedied:

- corrosion due to the mixture's chemical aggressiveness
- the design of the heat box and the fit to a suitable type of focussing reflector
- the elimination of phase-separation problems in long-term operation
- sundry technical problems and
- ascertainment of the system's acceptance potential.

To the knowledge of the authors, there is presently no work being done anywhere on a solar cooker featuring latent-heat storage.

All heat-accumulating types of solar cookers have one thing in common: they are so expensive as to be unaffordable for the average family in a developing country. Consequently, it is precisely the rural population that loses out. With that in mind, two cooker designers sized their units large enough to serve at the institutional level, i.e. in schools, hospitals, etc.

Some of the <u>advantages</u> of heat-accumulating solar cookers are:

- Cooking can also be done in the evening, i.e. after sunset
- The cook needn't stand out in the sun
- The food is easily accessible
- Some units can be used for baking, too
- The achievable temperatures are higher than those produced by cooking boxes
- Some collectors require little or no solar tracking.

Their <u>drawbacks</u>:

- Their extremely high cost (up to US$ 2000 or more) could only be lowered by mass production
- Various "exotic" materials and components are used
- The storage units are heavy and immobile
- It is hard to get a good, uniform transfer of heat from the heat store to the cooking pot; one such cooker requires the use of pots with surface-ground bottoms (electric cooking pots).

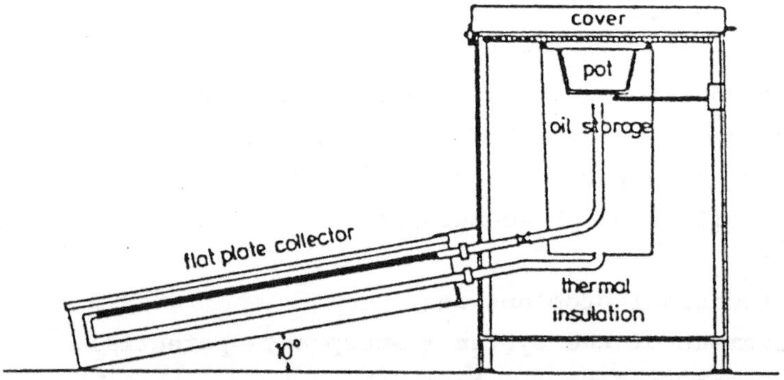

<u>Figure 7:</u> Solar cookstove with integral oil storage by ISE

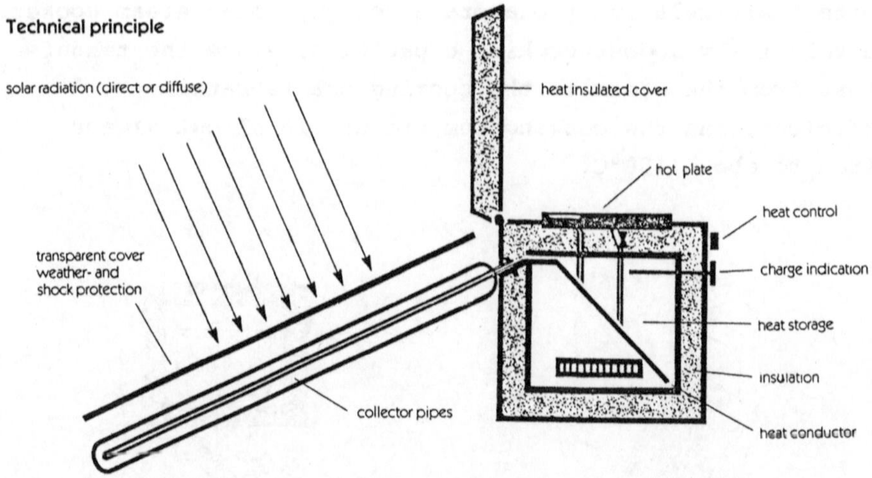

Figure 8: Heat-pipe storage cooker by Pöhlmann/Stoy; also available: an alternative version with a domed lid that closes on the cooking pot

2.1.4 Steam cookers

Solar steam cookers use an efficient (flat plate) collector to generate steam, which then rises to an elevated cooking box, where it heats the bottom and sides of the cooking pot and, hence, the food.

Fixed-focus concentrators, heat-accumulating solar cookers and other designs in which the collector is separate from the stove offer the potential advantage of being able to cook in the shade or even indoors. The cooking temperatures are moderate, and the efficiency is very low (about 15%), with accordingly long cooking times; such cookers are quite elaborate and expensive in relation to their performance.

In addition to the advantage of being able to cook in the shade or indoors (thanks to separation of the collector from the stove), steam cookers also afford easy access to the food, the flat plate collector makes use of diffuse radiation, and the unit rarely or never has to be aligned with the sun.

Despite their relatively elaborate design, solar steam cookers have relatively modest cooking capacities, since the transfer of heat from the steam to the cooking pot is rather inefficient, and the cooking temperature is almost always limited to about 100°C.

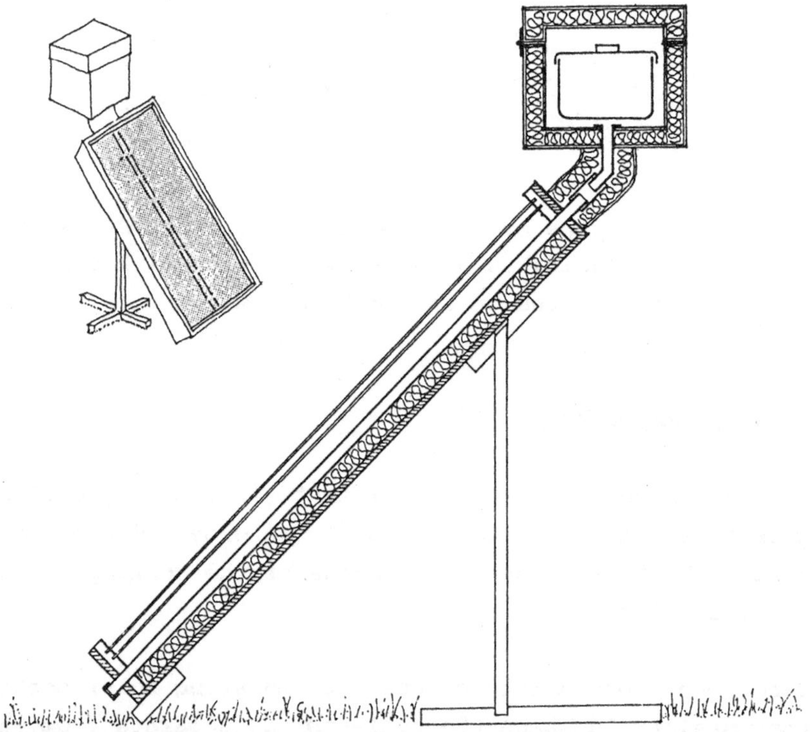

Figure 9: Solar steam cooker developed by the BRACE Institute

2.2 Function

Solar cookers/stoves absorb solar energy and convert it to heat for cooking or baking.

2.2.1 Solar radiation energy

Solar radiation is electromagnetic radiation in the 0.28...3.0 μm wavelength range. The solar spectrum includes a small share of ultraviolet radiation (0.28...0.38 μm), the visible light range (0.38...0.78 μm) and infrared rays (0.78...3.0 μm), with the latter accounting for nearly half of the solar spectrum.

In the earth's atmosphere, solar radiation is received directly (direct radiation, S) and by diffusion in air, dust, water, etc., contained in the atmosphere (diffuse radiation, H). The sum of the two is referred to as global radiation, G:

$$G = S + H \quad (1)$$

On a clear day at high noon, the global irradiance can amount to as much as 1000 W/m^2 on a horizontal surface. Under very favorable conditions, even higher levels can occur.

The amount of incident energy per unit area and day depends on a number of factors, e.g.:
- latitude
- local climate
- season of the year
- inclination of the collecting surface in the direction of the sun.

The average annual global radiation impinging on a horizontal surface amounts to

approx. 1000 $kWh/(m^2a)$ in Central Europe, Central Asia, and Canada

approx. 1700 $kWh/(m^2a)$ in the Mediterranian and most equatorial regions approx. 2200 kWh/(m^2a) African, Oriental, and Australian desert areas (see fig. 14).

In general, seasonal differences in irradiation are considerable and must be taken into account for all solar energy applications.

Tilting the collecting surface some 30...50° to the South in the Northern Hemisphere or to the North in the Southern Hemisphere yields somewhat better wintertime results for the region in question, but also some losses in summer. In the tropics, a nearly horizontal receiving surface is generally most advantageous because of the sun's high altitude.

2.2.2 Transfer mechanisms

It takes several successive steps to transmit and convert solar energy from its reception to the point of actually heating food. Depending on the type of solar cooker being used, those steps differ from case to case:

Box-type solar cookers make use of both direct and diffuse radiation, with part of it being reflected off of the mirror lid and the remainder entering the cooking space directly through the transparent cover. Since practically everything in the box is black - the pot, its lid, the walls and floor of the box itself - most of the incoming radiation is absorbed. The black surfaces heat up and begin emitting longwave thermal radiation that is unable to escape through the glass cover. In a quasi-static state of radiation equilibrium, the pot also heats up and cooks the food by thermal conduction, with the aid of radiant heat from the lid. Additionally, steady circulation of the air in the box supplies more heat to the pot by convection, including from below, if the pot is not standing directly on the floor of the box. Transversal heat conduction through the floor of the box to the bottom of the pot (if the pot is standing directly on the floor of the box) has a similar though somewhat less pronounced effect.

Reflector cookers exploit only the insulation that impinges directly onto the reflecting mirrors, which redirect it toward the cooking pot. The dark-colored bottom and sides of the pot absorb the radiation and transfer the heat to the food. The concentrated radiation has a higher power density and is therefore able to generate higher temperatures and temperature gradients at and within the pot. While that does make for faster cooking, it also makes the food more susceptible to burning than would be the case in a solar cooking box.

Heat-accumulating solar cookers store heat in a solid or liquid medium. In a simple case, e.g. a cooking box equipped for heat storage, the process of heating up the storage mass can be exactly the same as for heating food: insulation, absorption, thermal radiation, convection, heat conduction. The food (in a pot in or on the heat store) gets most of its heat by conduction and radiation from the heat store. The more advanced types of heat-accumulating solar cookers have very elaborate heat transfer mechanisms. They collect solar energy not directly in the cooking containment, but in highly efficient tube/flat-plate collectors. The heat is transferred at a relatively high temperature to the heat store by heat pipes or a thermal oil loop; in the former case a solid store is used, and in the latter case the heat is stored in a volume of liquid serving as an integral part of the collecting cycle.

The transfer of heat from the solid store to the hot cooking plate may occur directly by conduction or via an intermediate controllable heat pipe. In either case, the hot plate imparts heat to the pot by heat conduction (assuming good contact between the two).

The liquid heat store may include a depression that serves as the cooking containment, in which case both the storage medium and the air between the containment wall and the cooking pot give rise to lively convection. Together with thermal radiation from the walls of the containment and heat conduction into the pot from below, this arrangement provides for efficient heating of food.

2.2.3 Loss mechanisms

Not all of the energy irradiated on a solar cooker can actually contribute directly to the cooking process by heating the food and/or boiling the water. The losses occurring along the way are due mainly to:

- optical radiation ($\lambda < 3\ \mu m$) through partial reflection from and absorption by the cover (glass) in combination with incomplete absorption inside the cooking box/pot, with the remainder being reflected;
- thermal radiation (long-wave, $\lambda > 3\ \mu m$) given off by all warm-to-hot parts of the solar cooker. Box-type solar cookers exploit the thermal radiation given off by the inside of the box and the cooking pot by trapping it under a glass cover that does not let the thermal radiation pass through, but instead absorbs it, heats up and then radiates heat itself;
- heat conduction, despite careful thermal insulation: through the insulation in the cooking box, through its transparent cover (two panes of glass) and, in the more elaborate versions, through the collector, the heat store and the pipes;
- convection by air, either circulating within the cooker due to differences in temperature at different points of the cooker, or due to the cooling effect of wind on the outside (and, to a certain extent, the inside) of the cooker.

With due consideration of the aforementioned transfer and loss mechanisms, the effective thermal power, Q'_{eff} (heat gain), inside a solar cooking box can be described by way of the following 4 terms:

i) optical gain through radiation, transmission and absorption (including reflection losses, since $\tau = 1$ and $\alpha = 1$):

$$Q'_{opt} = G' \cdot A \cdot \tau \cdot \alpha \tag{2}$$

Q'_{opt}	optical power gain
G'	irradiance
A	aperture of solar cooker
τ	transmittance through the cover
α	mean absorptance inside the cooker

ii) heat losses through the transparent cover, summarized by a coefficient of heat transmission for the aperture, U_A, including heat conduction, radiation and convection, and which, to be precise, is a function of temperature:

$$Q'_{L,A} = U_A (T) \cdot A \cdot \Delta T \tag{3}$$

$Q'_{L,A}$	thermal power loss through the aperture
U_A	heat transfer coefficient of cover (aperture)
T	temperature
ΔT	temperature head between the inside and outside of the cooking box

iii) heat losses through the bottom and sides; like for ii), but with a better (read: lower) heat transfer coefficient, U_W:

$$Q'_{L,W} = U_W \cdot (A+2bh+2dh) \cdot \Delta T \tag{4}$$

$Q'_{L,W}$	thermal power loss through floor and walls
U_W	heat transfer coefficient of the floor and walls
b	width
h	height
d	depth

(The dimensions of the bottom, b and d, are taken as approximately equal to those of the aperture, A.)

iv) convection losses through the joints of the cooking box. These depend on wind-induced differential pressures, the length of individual joints, l_i (particularly around the tiltable cover), the quality of the joint (coefficient of air permeability, a) and - instead of temperature - the enthalpy of the air: any moist warm air escaping from the box increases the total loss by the amount of its heat of vaporization:

$$Q'_{L,J} = V' \cdot \rho \cdot \Delta H' = a \cdot \rho \cdot \Delta H \sum_i l_i \cdot \Delta p_i^{2/3} \qquad (5)$$

$Q'_{L,J}$	thermal power loss through joints
V'	volumetric flow of air through the joints
ρ	density of air
$\Delta H'$	difference in the specific enthalpy of the air inside and outside of the box
a	coefficient of air permeability of joint
l_i	length of a joint
Δp_i	pressure drop across a joint

(From window engineering, we know that the volumetric flow of air is roughly proportional to $\Delta p^{2/3}$; accordingly the unit of a is defined as m3/(h m $Pa^{2/3}$).)

In sum, then, the effective thermal power Q'_{eff} of the solar cooking box, from equations (2) through (5), figures to:

$$Q'_{eff} = G' \cdot A = \tau \cdot \alpha - U_A \cdot A \cdot \Delta T - U_W (A+2bh+2dh) \Delta T - a \cdot \rho \cdot \Delta H' \cdot \sum_i l_i \cdot \Delta p_i^{2/3} \qquad (6)$$

This equation is similar to the one used for calculating the effective power of a flat plate collector, the difference being that the collector's comprehensive effective U-value has been replaced by two different heat transfer coefficients: U_A for the aperture and U_W for the walls and floor (this in order to do more justice to the given geometry = box), and that the losses have been supplemented by joint convection (5) - a very important term for cooking boxes.

<u>Reflector cookers</u> usually show especially high radiant and convective heat losses, since the cooking pot is exposed to the effects of wind and radiation imbalance with respect to the surroundings. Add to that the optical losses attributable to less-than-ideal reflection by the mirror (partial absorption,

diffuse reflection and geometrical defects in the mirror) and imperfect absorption by the cooking pot.

2.2.4 Efficiency

As far as solar cookers are concerned, the term efficiency is not dealt with uniformly in professional circles. Different experts take different degrees of liberty in defining the term, and the results are accordingly open to debate. A recent proposal of definition and experimental evaluation of solar cooking box efficiency are described in /180/.

Defining the efficiency, η, as the ratio between effective thermal power and the incident radiant power (global irradiance x area of aperture),

$$\eta = \frac{Q'_{eff}}{A \cdot G'} \tag{7}$$

the efficiency equation emerges from equation (6) as follows:

$$\eta = \tau \cdot \alpha - [U_A + U_W (1 + \frac{2h(b+d)}{A})] \cdot \frac{\Delta T}{G'} - \frac{a \rho \Delta H'}{AG'} \sum_i l_i \cdot \Delta p_i^{2/3} \tag{8}$$

As in the conventional equation for flat plate collectors, this equation covers the optical efficiency, an approximately linear loss term as a function of the reduced differential temperature $\Delta T/G'$ with two coefficients of heat transmission, U_A and U_W, both of which, in second approximation, may also be regarded as temperature-dependent, supplemented by an enthalpy-dependent loss term accounting for joint convection.

A mean <u>empirical</u> efficiency can also be defined for non-accumulating solar cookers with little heat capacity: the ratio of useful heat yield divided by the global radiation passing through the aperture in the same time interval, with constant irradiance, applied to the process of heating 1 l of water from 20°C to 100°C. This efficiency term averages out the various operating states of the heating process, during which the instantaneous efficiency is subject to change: for cooking boxes, as for flat plate collectors, the instantaneous efficiency decreases with increasing temperature and vice versa.

$$\eta = \frac{m \,.\, c \,.\, \Delta T'}{G \,.\, A} \tag{9}$$

m mass (here: 1 kg water)
c specific heat capacity (1.19 kJ/kg K)
$\Delta T'$ temperature interval of the heating process

G is the global radiation, i.e. the integral of irradiance, G', passing through the aperture plane during the time it takes to heat the water from 20°C to 100°C:

$$G = \int_{t_{20}}^{t_{100}} G' dt \tag{10}$$

Capacitive effects (e.g. preheating of the solar heater prior to testing, thermal mass of the cooking pot, ...) can influence the results, as can wind and (for some types of cookers) the angle of radiation incidence relative to the aperture, in addition to variations in irradiance. The strong point of this definition of efficiency lies in the simplicity of its experimental realization.

No such efficiency values can be stated for heat-accumulating solar cookers, at best for the heat-collecting components. Those, however, would be pure collector-efficiency values that would leave unaccounted for the heat storage losses and those related to the heat transfer from the store to the food.

2.2.5 Cooking and baking: temperatures and performance

Cooking is the process of treating food with heat in order to enhance its taste and nutritional value and/or neutralize certain ingredients /71/.

During the cooking process, the cell structure is loosened and, in part, broken down, thus allowing more efficient use of the nutritive substances. At the same time, certain aromatics and flavoring substances can form, thus stimulating a person's appetite and causing increased secretion of digestive juices. With a view to keeping the loss of heat-sensitive vitamins and amino acids within limits, the cooking temperature and duration should not exceed what is necessary /71/. The extent to which overcooking and/or extended warming of food, specially in aluminum vessels, may lead to the formation of toxic substances in some dishes should be investigated separately /113/.

Some types of solar cookers are good for certain kinds of cooking processes, and others are good for others. Reflector cookers, for example, reach such high energy flow densities - and correspondingly high temperatures (over 100°C) - that those of appropriate design can be used for frying, roasting and barbecueing - or even baking crispy bread and cake. The problem with such high temperatures, of course, is that the food might burn. The same applies by analogy to heat-accumulating solar cookers.

Box-type solar cookers, by contrast, have a concentration factor of somewhere between 1 and 2 (with a reflecting lid) and a correspondingly low energy flow density. Consequently, they do not generate such high temperatures, i.e. most of the cooking is done in the sub-100°C temperature range.

For most kinds of food, though, that much heat is thoroughly adequate: protein (e.g. in meat) begins to undergo denaturation at about 45°C; the accelerated swelling of rice, lentils, etc. sets in at roughly 80°C, and the fibrocellular structure of both meat and vegetables is usually destroyed at temperatures of less than 100°C.

The main reason why cooking is preferentially and effectively accomplished at the temperature of boiling water (at atmospheric pressure) or at higher temperatures (e.g. in pressure cookers) is that the high temperature gradient, i.e. a large difference between the temperature inside of the food and the temperature of its surroundings, helps the heat penetrate into the food more quickly; lower temperatures would usually suffice to induce swelling and disintegration of the fibrocellular structure - but "low-temperature cooking" would naturally take longer, too.

The same applies to baking. Many kinds of dough bake well at less than 100°C. Here, too, a relatively low baking temperature, correspondingly small differential temperature and accordingly poor transfer of heat into the bread or cake would make the baking process take much longer. In addition, the substantially higher temperatures needed to give bread a crispy crust can only be provided by concentrating-type solar ovens and some, but not all, heat-accumulating solar cookers.

The <u>thermal power</u> needed for cooking serves two purposes:

- for heating the food to cooking temperature: This requires a relatively high thermal power input (by comparison, an electric range would be set at 1500...2000 W). Reflector cookers/concentrators and some types of heat-accumulating solar cookers (fully charged) are able to develop such high power levels, while box-type solar cookers cannot. The latter therefore take longer to complete the hetup process. The time expenditure can be shortened by placing some sort of heat-storing element (dark-painted rocks, bricks or the like) in the solar cooking box.
- for keeping the food warm or maintaining the cooking temperature: The heat requirement of an ideally insulated unit can approach zero for this function, which corresponds to that of the old fireless cooker (haybox); by comparison, an uninsulated pot on an electric range would need about 900 W without a lid and only about 300 W with a lid (presupposing good heat transfer to the bottom of the pot). Putting the same pot in a fully insulated cooking containment (solar

cooking box, fireless cooker) further reduces the power requirement, which is primarily a function of the box's heat losses. As long as adequate thermal power is available, all watery dishes eventually stabilize at a temperature of roughly 100°C; the excess power is dissipated in the form of heat of evaporation as the water in the food gradually evaporates.

2.2.6 Thermal output

The thermal output of a solar cooker is determined by the solar irradiance level (max. 1000 W/m^2), the cooker's effective collecting area (usually between 0.25 m^2 and 2 m^2), and its thermal efficiency (usually between 20% and 50%). Table 2 compares some typical area, efficiency and cooking-power values for a box-type solar cooker and a concentrator.

Table 2: Standard values for area, efficiency and power output of reflector cookers and cooking boxes

	Area (m^2)	Normal efficiency	Output W at 850 W/m^2	Time needed to cook 1 liter of water
Reflector cooker	1.25	30 %	320	17 min
Cooking box	0.25	40 %	85	64 min

As a rule, reflector cookers have a much larger collecting area than do cooking boxes. Consequently, they are able to generate a much higher power output, meaning that they can boil more water, cook more food, or process comparable amounts in less time. On the other hand, their thermal efficiency is lower, because the cooking pot is completely exposed to the cooling effects of the surrounding atmosphere.

In many tropical and subtropical countries, one can count on clear skies and normal daily insolation patterns for most of the year. At about midday, when the global radiation reaches a good 1000 W/m^2, the table-2 power levels (~ 50...350 W, depending on the type and size of the cooker) may be regarded as quite realistic. The solar irradiance is naturally lower during the morning and afternoon hours and cannot be fully compensated for by solar tracking. (Note in that connection that box-type solar cookers also can be kept more or less in line with the sun.)

By way of comparison: burning 1 kg of dry wood in one hour yields approximately 5000 W times the thermal efficiency of the cooking facility (15% for a three-stone hearth and 25-30% for an improved cookstove). The thermal power actually reaching the cooking pot therefore amounts to between 750 and 1500 W.

Figure 10 shows the typical time history of global irradiance from sunrise to sunset, with the lower curve indicating the diffuse radiation. Reflector cookers cannot exploit the latter, since they are only able to focus the direct radiation, but solar cooking boxes and cookers with non-focussing collectors can.

Solar irradiance drops off sharply under cloud and during the rainy season. The lack of direct radiation leaves reflector cookers without the slightest chance, and cooking boxes can do little more than keep prepared food warm. Which brings us to the intrinsic weak point of solar cooking - no matter what kind of device is used: on cloudy and rainy days - adding up to between 2 and 4 months per year in most Third World countries - cooking has to be done according to conventional methods, e.g. over a wood/dung fire or on a gas/kerosene-fueled cooker.

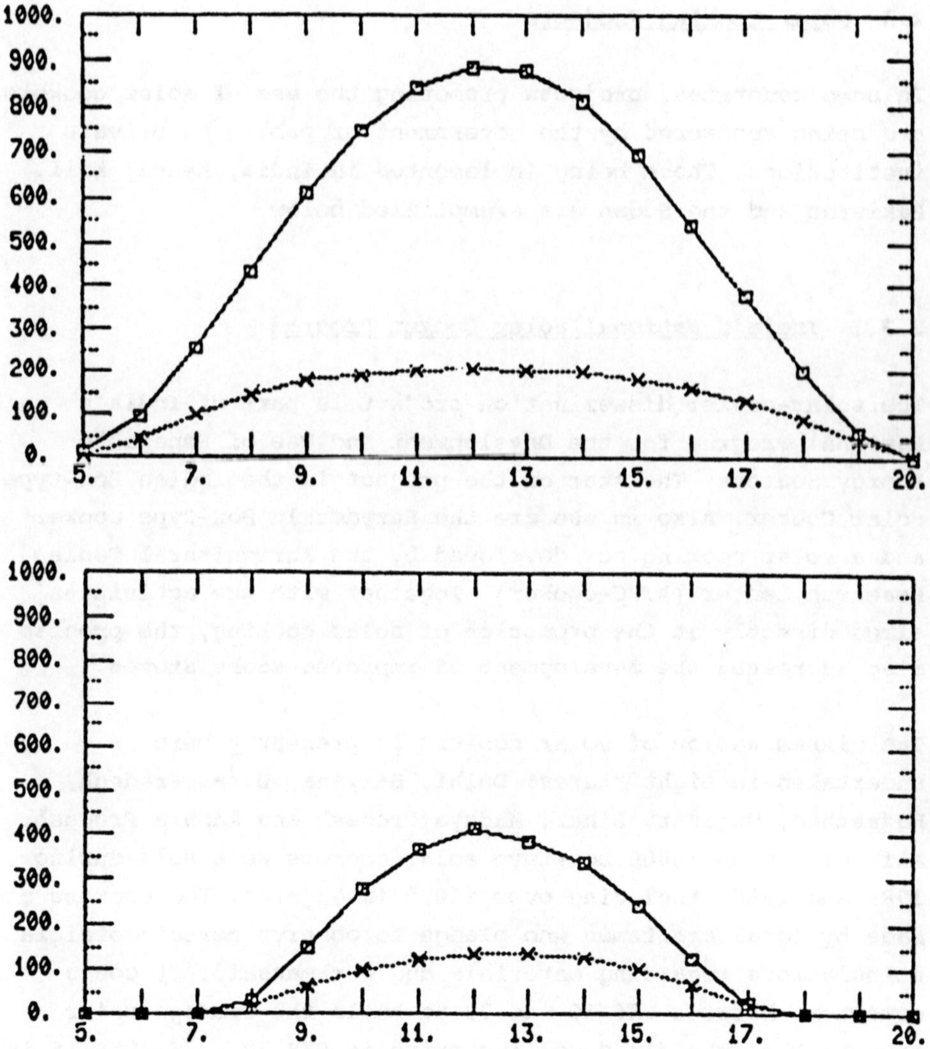

__Figure 10:__ Global (-□-□-) and diffuse irradiance (-x-x-), measured in W/m^2, vs. time for an average day in June (top right) and an average day in December (bottom left) on the southern Mediterranean coast /59/

2.3 Solar Cooking Projects

In some countries, projects promoting the use of solar cookers are being sponsored by the government or public or private institutions. Those being implemented in India, Kenya, Mali, Pakistan and the Sudan are exemplified below.

2.3.1 India's National Solar Cooker Program

The solar-cooker dissemination project is part of India's National Program for the Development and Use of Renewable Energy Sources. The star of the project is the Indian Box-type Solar Cooker. Also in use are the Suryamuklu Box-type Cooker and a solar cooking box developed by the Agricultural Tools Research Center (ATRC-cooker). Together with the activities aimed directly at the promotion of solar cooking, the program also addresses the development of improved adobe stoves.

The dissemination of solar cookers is presently being undertaken in eight states: Delhi, Haryana, Uttar Pradesh, Rajasthan, Gujarat, Bihar, Madhya Pradesh and Andhra Pradesh. All told, some 70000 box-tpye solar cookers were sold during 1985 and 1986, including over 21000 in Gujarat. The cookers are made by local craftsmen who pledge to observe certain official stipulations regarding materials and workmanship. It costs somewhere between US$ 60 and 70 to build the average solar cooker. The subsidized selling price is US$ 30...35 (US$ 15 in low-income regions). The following information applies mainly to the two states Gujarat and Delhi.

Target group: The solar cooker dissemination program is presently targeted on families with medium and above-average incomes. In urban areas this means successful shopkeepers, craftsmen and senior civil servants. In rural areas, those involved in the project are also salaried employees, prosperous landowners and cattle farmers. Members of this group earn between US$ 15 and US$ 200 per month. In addition, the Indian army has ordered 300 solar cookers. The lower social strata cannot afford to buy such cookers.

The target groups in Gujarat and Delhi use mostly wood, propane, kerosene, charcoal and agricultural residue as fuel for cooking. Due to limited resources, commercial energy carriers like propane and kerosene are routinely rationed. Consequently, most women use whatever combination of fuels they can get. Between 10% and 20% of the monthly family income is spent on fuel.

Dissemination program: The gradual diffusion of solar cookers is being promoted with the aid of a large-scale publicity program. At the supraregional level, the advantages of solar cooking are expounded on television, on the radio, in newspapers and in periodicals. Additionally, numerous craftsmen, salespeople, institutions and development organizations are participating in the solar cooker dissemination drive. At the village level, solar cooking demonstrations are held at village meetings and markets and in schools, hospitals and other public-access buildings. At dancing competitions, for example, the winning couple gets a solar cooker. In urban areas, solar cookers are demonstrated and sold in shopping centers and at seminars and exhibitions.

Also, all kinds of compainies recently have begun to participate in the dissemination program by offering their employees a US$ 10-premium toward the purchase of a solar cooker. Other firms pass solar cookers out to interested parties in return for a small monthly payment. Some coal miners are being given free solar cookers instead of free coal.
Each solar cooker comes with a pamphlet explaining its proper function and operation.
Basically, the dissemination porgram embraces two approaches:
- convincing the target group through personal experience (seeing, tasting, feeling)
- enlisting the aid of influential (opinion-forming) individuals and groups to help motivate women.

According to the first approach, women are expected to attend cooking demonstrations to "experience" how solar cookers work by doing some experimental cooking themselves. In Gujarat, solar cookers are loaned out to families interested in finding out what solar cooking is all about.

The other approach assumes that women will be quicker to accept solar cookers, if they have the opportunity to see them being used by influential, prominent or highly educated individuals or groups. In other words, if it becomes known that members of the upper social strata have taken to using solar cookers, other - lower - social strata should follow suit. The presumption here, of course, is that low-income families will presumably try to "keep up with the Jones's". The initiators therefore push the program with posters in which solar cookers are depicted as a symbolic part of a better way of life. The aim is to give solar cookers such a good image that the potential buyer views them in the same light as any other luxury article like a radio, television or refrigerator.

Figure 11: Advertising poster for solar cookers in Gujarat, India

In contrast to other government agencies, the Gujarat Energy Development Agency (GEDA) does not measure the success of the dissemination project against the number of solar cookers sold, but against the number of those actually in use. That is, the promotion campain is not supposed to merely help sell the cookers, but should also and simultaneously appeal to people's curiosity - get them to ask questions and try it themselves.

There is at present no regular backstopping being provided to families who already have a solar cooker, though such a program is planned for the near future in Gujarat. Systematic studies of the basic socioeconomic situation of the target group are also planned.

The heavily subsidized, state-controlled sale of more than 70000 solar cooking boxes all over the country makes India the only country (possibly outside of China) in which solar cookers have achieved any appreciable degree of diffusion. This is probably attributable to a combination of governmental efforts, particularly at the state level, and the technical character of the cookers themselves. At any rate, just how durable that success will be remains to be seen. Box-type solar cookers appear to be the only approach to solar cooking that has gained nominal - if conditional - acceptance. But the relatively large number of units sold should not be allowed to obscure the fact that those with a solar cooker constitute only a tiny minority of India's 764 million inhabitants. Despite the genuine merits of box-type solar cookers, and despite the large-scale publicity campaigns in their favor, no massive dissemination of such cookers has been achieved to date.

2.3.2 Indo-German Dhauladhar Project

Measures taken to promote the testing and introduction of solar cookers in the northern state of Himachal Pradesh are part of an overall scheme aimed at achieving ecological stability in the erosion-imperiled project area. The Indo-German Dhauladhar Project (IGDP) includes erosion-prevention and afforestation programs, activities designed to improve crop yields and animal production, and the introduction of energy-saving adobe stoves (Dhauladhar Chulha). Work on the development of solar cooking equipment began in 1985. By 1986 some 20 solar cookers were built and dispensed. Existing plans called for a doubling of the number of solar cookers in use by year's-end 1986. The "Dhauladhar Solar Cooker" is a stationary, partly home-built, box-type solar cooker made of adobe; the requisite material costs US$ 4...10, depending on the desired model.

Target group: The target group in this case is the general populace in the project area, about 12% and 23% of which belong to the highest and lowest castes, respectively. Some 90% of the project-area inhabitants are farmers. Nearly 40% of the families do sharecropping on at least part of their land, and 39% avail themselves of seasonal labor. More than 80% own cattle (cows, sheep and goats). Nearly all have some form of supplementary income from a family member in military or civil service.

Wood is relied on for most cooking and heating; about 10% of the families buy their firewood, while the other 90% spend two or three hours daily gathering it. Kerosene is used in about every other household to cover a small fraction of the total fuel requirement. If all cooking and heating were to be done with kerosene, consumption would run at about 13 l per month and family, at a cost of roughly US$ 13. Agricultural waste products like cow manure, rice straw or pine cones play a lesser role as household energy sources, being used mainly during the rainy season due to a shortage of dry wood.
All women in the project area do their cooking on a traditional indoor Chulha made of adobe clay. A second means of cooking - mainly for tea - is sometimes available in the form of a kerosene stove.

Dissemination program: Once some initial positive experience had been gathered in the course of the technical shakedown phase, solar cookers were installed in the homes of potential promotors, i.e. field workers, teachers, women's group chairwomen, etc.. The local craftsmen are familiarized with the production techniques by way of on-the-job training during construction of the stove. From October 1986 to September 1987, 133 solar cookers of the stationary Dhauladhar have been installed.

Meanwhile it turned out that the cooker is only accepted to a limited degree. The limited acceptance is considered to be due to climatic constraints and the firewood situation: It is assumed that the solar cooker is principally accepted in those areas where only commercial fuel is available. For this reason, it is intended to disseminate further solar cookers mainly in those villages which have no free access to firewood.

In 1988, a 6-page information/advertisement paper has been prepared for decision makers and/or for potential users of the Dhauladhar s.c., more printed information is currently being prepared.

2.3.3 ATDO Project in Pakistan

The Appropriate Technology Development Organization (ATDO), organized within the Ministry of Science and Technology in Karachi, began building solar cookers in early 1986. The ATDO cooker is described in the Appendix. The project objective is to test and disseminate solar cookers in the western regions of the country, where wood is scarce. The organization is also involved in the development of wood-saving clay stoves. The solar-cooker program; which represents only part of ATDO's numerous activities, is presently in its shakedown phase. Since work began, between 15 and 20 box-type solar cookers have been produced at a cost of US$ 24 apiece and sold for US$ 30 apiece. The project is being funded by the competent ministry.

Target group: Up to now most project activities have been concentrating on the Greater Karachi Area. The target group comprises independent tradespeople, middle-income salaried employees and others who can afford to buy a solar cooker at unsubsidized prices. The average monthly income of the target group amounts to US$ 85...90, of which roughly 8% is spent on fuel.
The project activities were recently extended to the rural areas of Sind Province, where the main thrust is being directed at the wood-starved Thar Desert, where most of the people work on the farming estates of major landowners. A few manage to produce some extra food that can be sold to supplement the average farm worker's monthly income of US$ 13...17. Due to the scarcity of wood, about 30% of the housewives fuel their fires with dry cow pats.

Dissemination program: ATDO's involvement in the production of solar cookers began in early 1986. Accordingly, the project is still in its initial phase, i.e. testing is still more important than mass dissemination. The introduction of solar cookers in rural Sind is being closely coordinated with local production centers, national development organizations and field workers from the United Nations Educational, Scientific and Cultural Organization (UNESCO). Some 500 villages have been selected to participate in field testing of the solar cookers. In contrast to the urban dissemination concept, the rural women are supposed to be advised and informed less individually, i.e. more within the framework of communal events. In that connection, the organization relies heavily on the female staff of the women's education centers (WEC's) for providing information and advice. The solar cookers are demonstrated at the WEC's. In addition, the women are given pamphlets containing further information. Any woman participating in such events is expected to become a multiplier within her own household environs. With deference to anticipated acceptance problems, ATDO is giving its activities a two-prong thrust, namely solar cookers and fuel-saving adobe stoves, the handling characteristics and function of which are more in line with the women's traditional cooking habits.

Gradual expansion of the project activities into regions affected by a distinct shortage of fuels would certainly be worthwhile, since the broader base would enable a more comprehensive assessment of the real acceptance potential for solar cookers among the general populace. Incorporating such activities into existing organizational structures (women's education centers, etc.) and cooperation with the local staff of various institutions/agencies will not only yield positive results, but also force alteration of the previous dissemination strategy, which addressed only individual households. In future, solar cookers will be presented and demonstrated at joint-effort gatherings and seminars for women.

2.3.4 Orangi Project in Pakistan

The Orangi Pilot Project is sponsored by a nongovernmental organization operating in Orangi Town, a Karachi Suburb. Alongside of numerous self-help activities and income-generating projects, the use of solar cooking boxes was added to the scheme in 1986. By the end of that year, about 30 solar cookers had been built and demonstrated to 350 families. The cooker in question is a rectangular box-type solar cooker with a reflecting lid. Due to lack of interest on the part of the local population - not a single family was willing to test-use a solar cooker for a prolonged period of time - the solar-cooker activities are being discontinued. The project staff attributes the lack of acceptance not so much to financial aspects, since part of the local populace has regular income - as to the technical inadequacy of the cooker itself and to the women's lack of willingness to alter their traditional cooking habits. As a result, the organization is now attempting to introduce wood-saving adobe and metal stoves. A major share of the solar cookers built to date were sent up north to the mountainous Gilgit region for field testing; to the author's knowledge, no evaluation has been conducted yet.

2.3.5 SERVE Solar Cooker Project in Peshawar, Pakistan

SERVE (Serving Emergency Relief and Vocational Enterprises), a voluntary agency based in England, initiated a pilot project to test the acceptability of solar cookers among Afghan refugees in Northern Pakistan in 1984. In 1985, a solar cooker dissemination project was started, which is still going on /184/.

Target group: Since the invasion of Afghanistan in 1979, over three-million Afghans have fled to Pakistan. In and around the refugee camps, there is a severe shortage of firewood and other natural fuels, and the ongoing fuelwood need and consumption of the Afghans has resulted in threatening the ecology of the region and gives rise to serious tensions between Afghan and Pakistani people.

A 1983 survey in the Northwest Frontier Province of Pakistan /184/ showed that 88% of the refugees burnt wood (approx. 0.85 kg per person and day), and 83% of them used kerosene (appr. 1 l per day and family). Three man-hours were spent per day and family for fuel collection. The refugees have seen cooking fuel as one of their greatest needs.

Most of the refugee women are illiterate, which is important with regard to their capability and willingness to adapt their cooking and eating habits to the use of solar cookers.

Pilot Project: In 1984, a square box fiberglass cooker with mirror lid was introduced. 2240 items of this type have been produced and distributed. Over 80% of the families in the project were reported to use the ovens on a regular basis /184/.

Altogether 16 models of different solar cookers were tested. Parabolic concentrators were judged not to be suitable for the needs of Afghan refugees.

Model Change, Production, and Marketing: The oven went through a number of design changes between the completion of the pilot project and the actual beginning of production in mid 1985. The square box shape was replaced by a rectangular shape. The interior oven liner and pots were painted with a high-temperature, non-glossy black paint. Handles and wheels made the oven easier to move and adjust. Each unit contained four 1 1/2 l pots, a pair of gloves, and illustrated cooking instructions.

Sales teams were formed to market the ovens in the refugee camps to display the cookers and prepared demonstration meals. Instructors taught refugee women how to use the cookers. The manufacturing price was US$ 56.00, the subsidized sale price US$ 18.00. Over 1780 cookers were sold from June 1985 to June 1987 /184/. Over 70% of the families are reported to use their solar oven consistently.

The rectangular 1985 solar oven model still had some drawbacks, and the following improvements have been suggested /185/:

- inclination of glass cover 20°...30°
- more distance between the cover glass panes
- clear glass with low or no iron content to minimize absorption
- more thermal insulation, 5...8cm thick
- more heigt in the inner box to obtain space for grid and convection distance under the pot
- Mylar foil reflector instead of glass mirror
- substitution of offset print aluminum sheet instead of iron sheet material for the inner box
- light wooden support instead of 4 metal wheels
- to replace the galvanized sheet metal casing by a fiberglass box with low weight and high life expectancy
- other measures in order to decrease weight and make the oven handy and transportable by one person
- to split the lower glass pane into two parts to relieve thermal stress and reduce breaking rate.

The 1987 Model: Most of these suggestions have been realized early in 1987, which led to a new SERVE Solar Oven model with less weight, more aperture, better performance, and higher reliability and lifetime which has a 20 K higher stagnation temperature. The manufacturing cost is about US$ 63.00, the sales price still US$ 18.00. The subsidy of $ 45.00 per oven is proviced by donors and international organizations like UNHCR, World Vision International, etc.

It was planned to extend the production of this new type, thus manufacturing 3700 ovens per year.

Conclusion: SERVE has produced and disseminated a solar cooker that performs satisfactorily and has been accepted by the refugees. This project is said to provide a model for introducing solar cooking to other segments of Pakistan society and in other parts of the world where similar needs and conditions exist /184/.

2.3.6 Solar cooker project in North Horr, Kenya

In cooperation with the Catholic mission in North Horr, the local production of a simple box-type solar cooker (ULOG cooker, tropical model 85, see appendix 1) began in 1985. The local carpenter and three helpers were shown how to build the cookers. The solar cookers, which are being built at the rate of roughly one per day, are sold at the prime-cost price of about US$ 25. Most of the > 40 units produced by end of 1986 had been sold, some of them were in service at the mission, and others were being used for demonstration purposes. All told, some 80 solar cooking boxes have been built since 1985 and sold to various missions and development organizations in northern and central Kenya, on the east coast and to the southeast of Nairobi. Most of the funding for the project is provided by the mission.

<u>Target group:</u> Basically, the target group consists of the approximately 3000 inhabitants of North Horr, some of whom are now-sedentary nomads (Gabra). The settlement is located on the fringe of Chalbi Desert, having appeared there long ago by reason of sweetwater springs. There are bushes in and around North Horr, and even trees grow along the riverbeds. The local inhabitants gather wood for their own use or for selling (10 branches for about US$ 10). The main sources of income are animal husbandry, handicrafts and odd jobs.

<u>Dissemination program:</u> The following activities are designed to inform and motivate the inhabitants of North Horr in order to promote the introduction and dissemination of solar cookers:

- exemplary use of the solar cooking boxes already in service at the settlement
- demonstrations at a central location
- solar-cooker demonstrations in various women's groups concerned with the production of palm-frond wickerwork. Those women are expected to convince other women of the advantages of solar cooking.

Also, individual household consultancy sessions were held for various groups of users. Local craftsmen are being taught how to build solar cooking boxes. Moreover, a three-day course in cooking-box construction was held for fifteen carpenter's apprentices at Marsabit Technical College.

The solar radiation conditions of the region (about 6.2 kWh/(m^2d) all year round) are favourable for solar cooking during the whole year (see radiation table of Lodwar, table 3). Nevertheless, users complain about too many cloudy days which aggravate a regular use of the solar cooker.

Some of the cookers had been given to several development organizations for field testing. However, no definite findings about the acceptance and use of the cookers, which had been distributed since 1985, could be established.

Big solar cooking box: In 1986, a big stationary box-type solar cooker was installed at the mission of North Horr. The 3.2 x 1.5 x 0.5 m box is made of bricks, the two 50 cm diam. flat cooking vessels are suspended at the level of the horizontal double glass cover, partially immerged into the box, where they are heated by convection and radiation from the absorbing walls of the inner box. This stationary box-type solar cooker is used to cook for 70 school children if weather conditions are good.

Fixed-focus hybrid solar cookers: From 1986 to 1988, three hybrid solar and wood fire cookers with 1, 7.6, and 22.4 m^2 apertures for 5, 54, and 500 l of food were constructed at a mission and two schools in Kenya. (For the fixed-focus principle see fig. 6; for details of these cookers see appendix 1).

2.3.7 GTZ project Sobako in Kenya

The project objective was to field test the Sobako 1 solar cooker - a linear-parabolic reflector cooker - with regard to:
- its appropriateness as a cooking appliance in an average household
- the anticipated savings on fuel
- improvement via design modification.

In addition, field trials were supposed to yield information on which to base suggestions for the future mass production of identical or similar cookers.

To that end, a solar cooker built in the Federal Republic of Germany was sent to Kenya. Two other cookers were produced by the Department of Mechanical Engineering, University of Nairobi, and by Nairobi-based Kenya Industrial Estate. The performance of all three solar cookers was investigated over a period of twelve months in various regions of Kenya. The target group to be familiarized with solar cooking comprised various low-income groups, primarily in rural areas.

It turned out that Sobako 1 and also its simplified successor model Soba 1 were complicated, high-cost, low performance solar cooking devices which were extremely unadapted to the user's needs (see description in appendix 1).

At the time of project implementation in 1977/78, it cost well over US$ 420 to build either a Sobako 1 or a Soba 1. Under assumed favorable conditions, the estimated cost price for economic lot sizes was expected to be about US$ 25. The calculated payback period based on fuel savings was 3.5 years; allowing for the money saved by home-baked bread shortened the calculated payback period to 1.8 years.

A six-month social-acceptance study was supposed to yield additional data. In that connection, eight additional solar cookers were to be built and demonstrated all over the country in cooperation with various governmental and non-governmental organizations. In addition, plans for producing and disseminating the cooker in large numbers were drawn up, and pertinent recommendations on a subsidy program were formulated for the government. Eventually, though, the severe shortcomings of the design principle led to discontinuation of all relevant activities. No methodical evaluation of acceptance was made that would satisfy even minimum criteria.

2.3.8 GTZ solar cooker project in Mali

Upon completion of a 1978 survey of existing solar cooking and baking appliances, the Munich-based Arbeitsgemeinschaft für Entwicklungsplanung (~ working group on development planning) received from GTZ a follow-up commission concerning the development and testing of solar cookers in Mali. To begin with, ten reflector cookers were built and tested in 1979/80 in cooperation with the Mali project partner. The cookers were field tested in two hospitals and four "Centres d'Animation Rurale" (CAR) from May through September 1980. The empirical data was evaluated in the form of a study entitled "Solare Koch- und Backgeräte - Prototypenbau und Erprobung" (~ solar cooking and baking devices - prototype construction and testing) /6/, and the following recommendations concerning the further development of solar cookers were made:

- appropriate long-term testing of solar cookers
- local production
- program for the broad-scale dissemination of solar cooking and baking devices.

Since the efforts aimed at introducing solar cookers were only moderately successful, the above recommendations were never implemented.

Target group: The "Comité Interservice pour les Energies Renouvelables" had eight solar cookers tested in four CARs (in Narena, Quelessebougou, Yangasso and Sorobasso - located 80 and 90 km to the south and southwest, and 400 to 450 km to the east of Bamako, respectively). A pair of cookers was given to two hospitals in San and Dioro to determine how well they would serve in heating water and sterilizing medical instruments.
A CAR is a rural training center in which young farmers can learn to read and write, take courses in modern agricultural production, and receive paramilitary training. Between 40 and 60 people live in the average CAR. Two women cook the meals for the trainees and their families. The two staple dishes eaten alternately for breakfast and supper are Tot and Couscous.

The women fuel their cooking fires primarily with wood, which the men gather between once and three times a week. Gathering wood takes between 2.5 and 8 hours, depending on the region. Only dead wood is used; indeed, the use of freshly felled wood is prohibited. The CAR women make tea on a portable charcoal stove, but cook the meals over an open fire (three-stone hearth). Solar cookers were introduced in the CARs and hospitals by way of demonstration and explanation of their function. The personnel received operating instructions and solar-cooker reporting sheets, but the users were given no trial-phase backstopping assistance.

2.3.9 <u>Sudanese Special Energy Program</u>

The use of solar energy for cooking is part of the Sudan's Special Energy Program, which took shape in the early 1980s and was promoted by the Kreditanstalt für Wiederaufbau (KfW, ~ reconstruction loan corporation) and the Deutsche Gesellschaft für Technische Zusammenarbeit (GTZ) GmbH. Some of the program's many planned and in part already implemented elements are afforestation, carbonization of cotton stalks, charcoal processing, energy-saving stoves, photovoltaic refrigerators in hospitals, solar-assisted educational television, windmills, and river pumps. The development and testing of a box-type solar cooker with reflecting lid, 80 of which were built, was more or less a spin-off from the mainstream project. Other types of solar cookers that have been tested in the Sudan, e.g. the axially parabolic reflector cooker Falco S/C and the Convective Solar Cooker, the latter being a rather large apparatus with convective solar collectors, have failed to gain acceptance or multiplication; no evaluations are known to have been conducted. According to a World Bank report, heating water is about the only use for solar energy that could find a market in the Sudan.

<u>Dissemination program:</u> As of this writing, prototypes have been distributed in 15 towns and given to interested institutions. Systematic testing is planned. The only nominal-scale test run as yet under controlled conditions was in Khartoum: relatives of staff members at the National Renewable Energy Research Institute (RERI) obtained the cookers at a special price of

about US$ 20 - compared to the going market price of US$ 40...80 - and used them for several months. Thirteen families participated in the test, and eleven families responded to a subsequent survey.

Target group: Even though the test families enjoyed incomes amounting to more than twice that of the average family, the heavily subsidized price was still too high - considering the cooker's meager performance. The potential users of the Sudanese solar cooker (if any) will most likely be limited to well-to-do urban households wishing to save charcoal.

Evaluation aspects: The makers of solar cooking boxes bemoan the lack of both a market analysis and a dissemination program with promotors, consumer surveys, media advertising and market strategy. One private producer has 14 cooking boxes waiting in vain for buyers. Observers report that the cooking boxes are of varying design and quality and that the product development phase is far from over; in other words, dissemination via the free market would still be premature, and the Ministry of Energy still does not have access to adequate dissemination channels. The test families tend to regard solar cookers as a supplementary means of cooking, mainly because the cooking times are still much too long. Other aspects are dealt with in other sections, e.g. chapter 3. At any rate, the solar cooker is not yet ripe for general diffusion; considering the given acceptance situation, it remains to be seen (and should be investigated) whether or not that ever will be the case.

2.3.10 The Chinese Solar Cooker Development and Dissemination Project

In September 1983, at the Conference of National Comparison and Exchange of Solar Cookers, 48 institutions presented 102 models of solar cookers, of which 80 had been tested. 41 of these had an efficiency above 60%, and 41 also had a thermal power above 1000 W. For most of them, the operating height was lower than 1.25 m. The specific cost ranged from 30 to 140 yuan/kW (US$ 8.00 - 38.00 per kW). Finally, the conference selected 24 models for dissemination because of their thermal efficiency, easy use, low cost, and adaptedness to the needs of a Chinese family of five /187/.

Tongging (Yongjing) County, in Gansu Province, started to popularize solar cookers already from 1978, and by end of 1983, there were about 20,000 solar cookers in use, 40% of the total number in the whole country.

As of this writing, more than 100,000 solar cookers have been in service in China. Almost all of them are of the eccentric axis concentrator type (see below).

<u>Cooking needs and model selection:</u> The preparation of chinese food comprises boiling, roasting, frying in boiling oil, etc. For these processes, high thermal power and high temperatures are needed. Solar cooking boxes, e.g. of the Indian type, cannot be used for many dishes/components of chinese food. For this reason, the solar cookers in use have mainly changed from the box type to the concentrating reflector type. The high thermal power is obtained by the choice of a large aperture (1...2.5 m^2), which results in 400...1200 W of thermal power. Thus, for some dishes, cooking on the solar cooker takes only half of the time that would be necessary on a wood fire. Maximum temperatures measured in the focal spot are - depending on the paraboloid precision and the concentration factor - 400...1000°C /187/.

<u>Fuel prices and savings:</u> Investigations in Taxang village have shown that a family of 5 persons consumes about 5 kg of firewood per day, which is 1825 kg per year. In a sunny region of China (2000 h/year of sunshine), a solar cooker can save 500...1000 kg of firewood per year. At the price of 0.20 yuan/kg (US$ 0.05/kg), this results in the saving of 100...200 yuan/year (US$ 30...60/year). "Practice has proven that in areas with better solar radiation and a serious lack of firewood, the solar cooker is feasible" /187/.

<u>Solar Cooker Technique:</u> Development of solar cookers in China started from the round parabolic dish. But with this design, it

is not possible to focus all radiation to the bottom of the pot, which results in low thermal efficiency, specially when the sun's altitude is low.

Using only a section of the lower part of the paraboloid allows the concentration of all radiation under the pot. This so-called eccentric axis concentrator has been developed to a large extent both in theory and practicee in the Energy Research Institute of the Henan Academy of Sciences at Zhengzhou. The eccentric axis concentrator type has become the most common type of solar cooker in China /186-188/.

The eccentric parabolic reflector was first made from conventional cement (160 kg), then from a thin layer of fiberglass reinforced cement (70 kg). If the glass fibers are cut short, they can be sprayed with the cement; this is suitable for industrial production.

Other cookers of this type have a wooden or metallic reflector shell. It can be folded once or twice, resulting in a handy transportable box. This so-called eccentric axis box-style solar cooker is still a pure reflector model and has nothing in common with other box-type models such as the Indian solar cooking box etc.

The reflecting surface was first realized by small pieces of mirror glass, then by a reflecting aluminum foil. But there have been problems concerning the quality and the stability of the reflecting foil surface.

2.3.11 Lessons to be drawn

No systematic evaluation of the above solar cooker projects was conducted prior to, during or after their implementation; the only relevant publications deal with India's national solar cooker program /116; 139/. Consequently, the lessons to be drawn from successful and unsuccessful projects must be of an extensively intuitive nature. The most immediate needs are:

- exact analyses of dietary, eating and cooking habits of the target groups, and which forms of energy they use
- more consideration of the target groups' buying power and "emergency strategies" for securing energy
- organization and implementation of training measures for local craftsmen
- regular quality controls at the place of manufacture in order to enable detection of hidden defects, e.g. forgotten insulation
- methodical field testing
- identification and exploitation of existing infrastructures and useful organizations (artisan groups, self-help groups, women's groups, etc.) for testing and disseminating solar cookers
- implementation of user sensitization measures designed to get across the importance of energy conservation
- including women in the project activities and realizing their importance as multipliers and promotors
- incorporating the project activities into existing women's programs (courses in dietetics, health & hygiene, handicrafts, etc.)
- familiarizing women with the utilization of such equipment
- volunteer participation in field-testing and/or dissemination programs
- provision of regular backstopping for all concerned, from the users to the makers, including any necessary follow-up studies
- making sure that the target group is the user/beneficiary and vise versa.

It would also be very good to promote back-up research. Most results of evaluation still rest more on anecdotes than on analyses.

2.4 State of the Art and Dissemination

Comparing prior surveys on solar cookers, e.g. /58/, with this study, one notices that the former make little mention of box-type solar cookers, while the present report is practically dominated by solar cooking boxes. Indeed, there seems to be a trend taking hold in the entire solar-cooking scence. While tens of thousands of box-type solar cookers are being produced - and sold - in India, other once dominant types of solar cookers appear to be losing favour. Consider for example the Fresnel reflector cooker by VITA, the reflector cooker from Mali, the Sobako, and the BRACE steam cooker. But in China, the majority of solar cookers in use are reflector type ones.

Initial successes in India and China are still infinitesimal in relation to the total population. The fact that "practically nobody in the world" is still cooking with solar energy, even though the advantages are quite obvious (and few are aware of the drawbacks), continues to spur the imaginations of engineers and solar experts, whose research data, inventions and new products - like the VIAX cooker and all four heat-accumulating solar cookers - must stand up to extremely stringent economic, technical and socio-cultural conditions of acceptance in the Third World.

The latest developments document the fact that inventors and engineers still have ample latitude for new ideas. They also confirm the observation that appropriate technology cannot be developed independently of the target group.

Solar cooking boxes have by no means reached the end of their technical evolutionary process, either. While the Telkes type (cooking box with a round floor, freely suspended pot holder and a funnel-shaped array of reflecting surfaces) was just plain too expensive, simple cooking boxes can now be made much more efficient at little or no extra expense. Using transparent insulation between the two panes of glass, for example, lets the sunlight enter practically without hindrance, but prevents the resultant heat from escaping through the cover. One such configuration is the transparent plastic honeycomb structure described in /145/.

New kinds of transparent insulation are being developed and tested /143/; they are expected to revolutionize the entire field of solar thermal technology and give new impetus to the spread of solar cooking, particularly in box-type solar cookers.

The usefulness of a solar cooking box can also be increased by including some means of internal heat storage, e.g. rocks. While that would mean that the cooker takes longer to heat up in the morning, it would also increase its primary cooking performance and help keep food warm until long after sunset. Such a simple and for all practical purposes free means of expanding the function of solar cooking boxes surely must have been tried out by various Third World users by now, but the authors have received no relevant reports. One possible drawback, namely the fact that adding a heat store to a solar cooking box would make it considerably heavier, still needs to be clarified.

__Figure 12:__ Box-type solar cookers with and without transparent insulation (honeycomb structure) in a comparison test (Photo: T.R. Ranganath)

India and China are the only countries in which efforts aimed at disseminating solar cookers have been at least partially successful. In India, where some 100,000 solar cookers have been sold to date (still minuscule in comparison with the total population), the "success" has been attributable to a massive dissemination campaign (cf. chp. 2.3.1).

In China, more than 100,000 solar cookers are already distributed and sold. Most of them are reflector cookers that develop the high temperatures needed to accommodate Chinese cooking habits. The authors do not know how many of these cookers are really in use at this time.

Whether or not solar cookers finally do gain widespread acceptance in some parts of the world remains to be seen. But one thing is for sure: wood and other natural fuels are becoming increasingly scarce in numerous developing countries with ample solar radiation; and fossil fuels can also be expected to become much more expensive in the not too distant future. Soon, most families in many areas will be finding no more natural fuels and will not be able to afford commercial fuels. Such families may find that solar cooking constitutes their only long-term alternative. The extent to which such arguments are being accepted by the target groups of solar cooker projects is still an open question - one that is examined more closely in the following chapter.

<u>Figure 13:</u> **Reflector cookers in China**

3. Conditions of Acceptance for Solar Cookers

Ultimately, the acceptance of solar cookers in and by the Third World depends on numerous prerequisites:

- available models: Is the solar cooker technically mature and of high quality?
- solar radiation: Is enough solar energy available?
- energy demand: Is the target population in a precarious energy situation?
- social situation: Are solar cookers acceptable from the standpoint of social structure?
- dietary patterns: Can solar cookers fit into the dietary patterns of the target groups?
- eating habits: Do the prevailing eating habits allow the use of solar cookers?
- cooking habits: Can the customary cooking techniques be preserved when using a solar cooker?
- cooking facilities: In what respect do the traditional cooking facilities differ substantially from solar cooking devices?
- technology: Which technical aspects of solar cookers are especially important in the social context?
- economy: Are solar cookers cost-efficient?
- side effects: What kind of overall impact, including side effects and consequential effects, do solar cookers have?
- additional conditions for acceptance: What other prerequisites must be fulfilled to ensure acceptance?

The above questions are defined and examined here in relation to the project results discussed in the preceeding chapter. This is followed by a questionnaire designed to facilitate evaluation of some major aspects of solar-cooker acceptance.

3.1 Preliminary Notes

Solar cookers are still at the development stage. No one model has yet achieved technical perfection; most prototypes are modified time and again.

Maturity: Most solar cooker projects consist more of prototype testing programs than of dissemination programs for technically mature solar cookers. Consequently, nonacceptance of an "unfinished" prototype does not necessarily mean that the future product - in the form of a technically mature, cost-efficient, series-built solar cooker - will also encounter similar nonacceptance and consequent failure of relevant dissemination efforts.

Quality: Most solar cookers, being prototypes, have not nearly reached their ultimate potential quality with regard to thermal efficiency, craftsmanship, stability, etc. The essential aspects of microquality and macroquality are dealt with in the following chapters (3.2 through 3.12).

3.2 Solar Radiation

The first and foremost prerequisite for success in a solar cooker project is adequate insolation, with only infrequent interruptions during the day and/or the year.

Insolation: The duration and intensity of solar radiation must suffice to allow the use of a solar cooker for prolonged, worthwhile regular periods. While cooking with solar energy is possible in Central Europe on a sunny summer day, a minimum irradiation of 1500 kWh/(m^2a) (corresponding to a mean daily insolation of 4 kWh/(m^2d) or 15 MJ/(m^2d)) should be available for any solar cooker project. Indeed, some sources speak of higher minimum insolation levels required.

But these annual data can sometimes be misleading. The essential condition for solar cooking is a reliable "summer weather", i.e. essentially predictable sequences of regular cloudless days.

Figure 14 /189/ shows the global distribution of annual irradiation in kWh/(m^2a). Visibly, the supply of solar energy varies substantially from country to country, even within the Third World's tropical belt. Thus, local data must be referred to - and they are not always available. Some examples:

- <u>India:</u> Solar radiation in most regions of India is good to very good for purposes of solar energy exploitation (see fig. 15). The yearly averages of daily annual global radiation range from 18 to 25 MJ/(m^2d), or 5 to 7 kWh/(m^2d), depending on the region. In most places, the insolation reaches its minimum during the monsoon season and is nearly as weak again during the months of December and January. In Delhi and Calcutta, though, the sunshine deficit during the monsoon season is not as pronounced as the one in winter. Figure 15 shows the regional distribution of global radiation incident upon India.
- <u>Kenya:</u> Kenya's climate and insolation potential are conducive to the use of solar cookers (see table 3). Kenya straddles the equator and therefore has a purely tropical climate. In Nairobi, the daily irradiation alternates between 3.5 kWh/(m^2d) in July and 6.5 kWh/(m^2d) in February, but it remains practically uniform (6.0...6.5 kWh/(m^2d)) in Lodwar, as indicated in table 3. Table 3 also shows that the solar irradiation in Nairobi is adequate for cooking with solar energy nine months a year (excluding June through August). On the other hand, conventional cooking facilities must be relied on for cloudy or hazy days. In the Lodwar area, though, solar cookers can be used year-round.
- <u>Mali:</u> The irradiation values vary from region to region but may be regarded as generally favorable. Irradiance increases steadily from south to north. Table 4 lists the 10-year averages of monthly sunshine duration, in hours, for 15 towns in Mali. The total annual sunshine duration comes to about 2,200 hours in the extreme southern parts of Mali and 3,500 hours and more in the country's northern reaches. An average value of about 3,000 sunshine hours per year is achieved around the 13th parallel, i.e. in the southern part of the country. The Narema and Quelessebougou areas get more precipitation and therefore have lower annual averages of sunshine duration (2,800 hours) than the rest of Mali.
- <u>Pakistan:</u> The solar radiation level in Pakistan decreases steadily from south to north, but remains substantially below India's average (cf. table 5). At the east-facing slopes of the mountains (Quetta, Peshawar) radiation is generally higher then in the lower part of the country. Minimum

irradiance in winter is conspicuous all over Pakistan. In many areas, there are 3 or 4 months with less than 4 kWh/(m^{2}d) of global irradiation (14.5 MJ/(m^{2}d)), which makes it very difficult, if not impossible, to use solar cookers.

- <u>The Sudan:</u> The Sudan gets a lot of sunshine; daily global radiation values measured horizontally range from 18.9 MJ/(m^{2}d) to 27.7 MJ/(m^{2}d) (corresponding to 5.3...7.7 kWh/(m^{2}d)) on an annual average.
- <u>China:</u> The total annual global radiation in China ranges from 3350 to 8400 MJ/(m^{2}a), which is a daily mean global irradiation of 9.1 to 23 MJ/(m^{2}a) or 2.5 to 6.4 kWh/(m^{2}d). The areas with more than 5900 MJ/(m^{2}a) (16 MJ/m^{2}d or 4.5 kWh/m^{2}d), cover more than 2/3 of the country /187/ (see fig. 15a).
- <u>Conclusions:</u> The above data underline the importance of various determining factors like climate, location, season of the year, etc. with regard to sunshine duration and solar intensity.

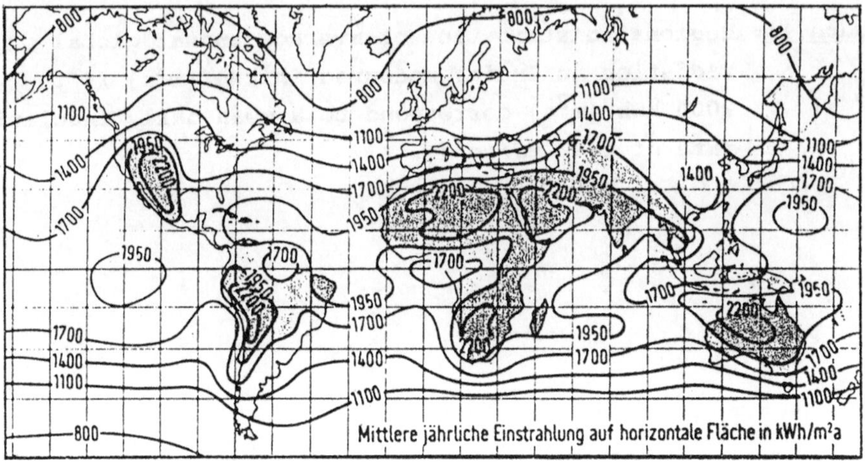

<u>Figure 14:</u> Mean annual irradiation on the horizontal plane in kWh/(m^{2}a) /189/.
2000 kWh/(m^{2}a) correspond to a mean daily radiation of 19.7 MJ/(m^{2}d) or 5.48 kWh/(m^{2}d).

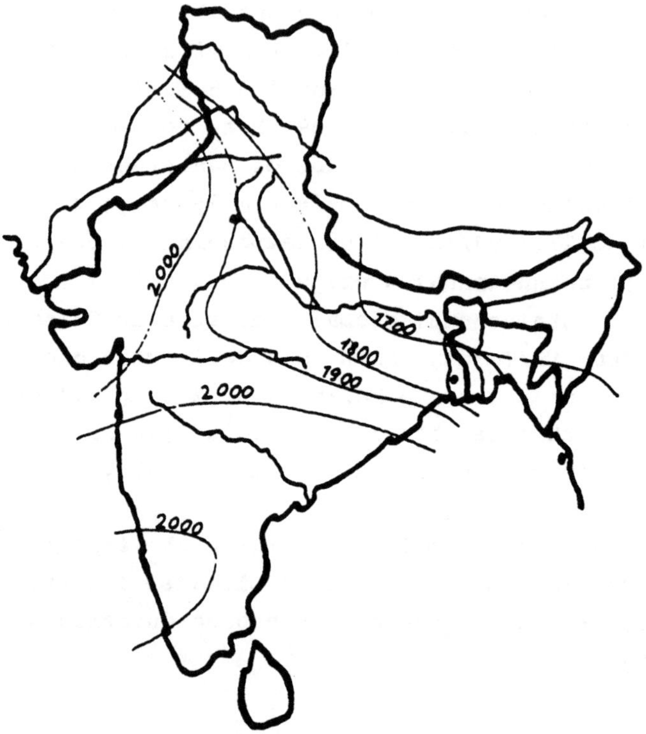

Figure 15: Regional distribution of average annual global radiation in India, measured in kWh/(m^2a) /6/. 2000 kWh/(m^2a) correspond to a mean daily insolation rate of 19.7 MJ/(m^2d).

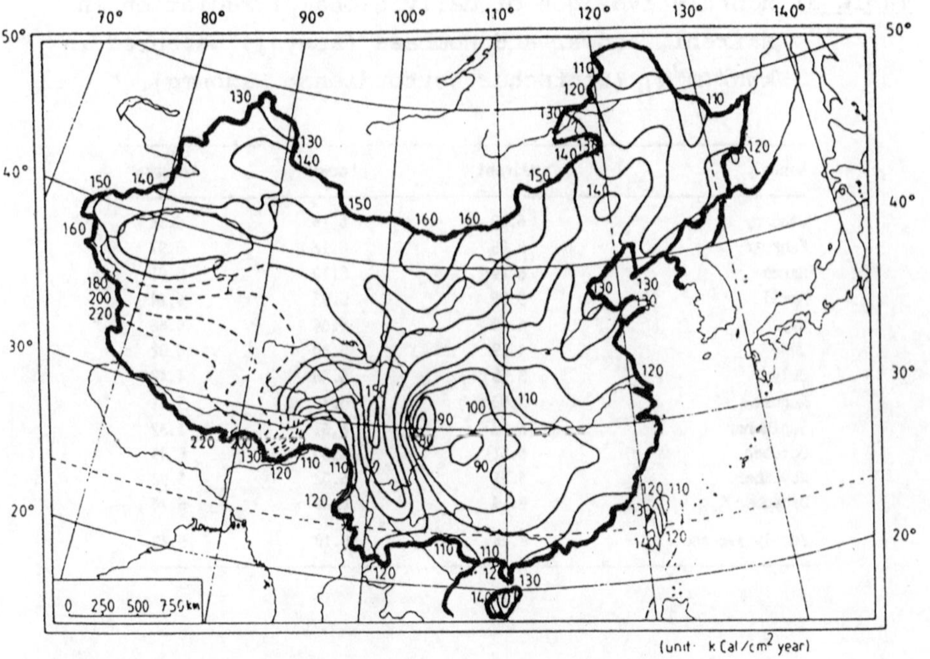

Figure 15a: Annual distribution of solar radiation in China in $kcal/(cm^2a)$ /187/.
$100\ kcal/(cm^2a) = 3.2\ kWh/(m^2d) = 11.4\ MJ/(m^2d)$ mean daily radiation.

Table 3: Monthly averages of daily global irradiation in Nairobi, Lodwar and Mombasa (Kenya), measured in $kWh/(m^2d)$ (Deutscher Wetterdienst Hamburg).

Month	Nairobi	Lodwar	Mombasa
January	6.30	6.16	5.84
February	6.55	6.16	5.97
March	6.19	6.10	6.01
April	5.25	6.03	5.24
May	4.64	6.34	4.48
June	4.19	6.10	6.02
July	3.59	5.86	4.48
August	3.93	6.30	5.01
September	5.28	6.51	5.52
October	5.61	6.30	5.49
November	5.31	6.02	5.92
December	6.13	6.29	5.76
Yearly average	5.25	6.18	5.48

Table 4: Mean monthly sunshine duration (h/month, 1961-1970) in Mali (Direction Nationale de la Meteorologie, Bamako, Mali)

Meteoro-logical station	Months												Annual average (h/a)
	J	F	M	A	M	J	J	A	S	O	N	D	
Bamako	272	256	280	247	246	238	203	178	208	254	261	261	2.903
Bougouni	277	240	241	212	238	231	206	174	183	238	259	265	2.764
Gao	295	274	293	279	307	272	294	280	279	294	291	276	3.432
Hombori	285	271	290	257	271	249	254	256	259	289	289	272	3.242
Kayes	266	247	295	293	276	242	217	194	229	258	262	265	3.044
Kenieba	263	258	294	291	272	240	187	155	185	235	258	264	2.900
Kidal	291	272	304	292	319	276	310	290	282	298	293	276	3.504
Kita	277	277	289	278	260	245	209	173	182	251	269	276	2.985
Menaka	268	261	271	242	263	254	264	255	257	181	277	278	3.171
Mopti	279	267	288	256	261	243	244	230	242	277	279	275	3.140
San	286	269	296	281	271	273	252	226	247	294	283	277	3.255
Segou	292	276	292	263	274	266	254	223	242	283	287	287	3.238
Sikasso	277	240	241	212	238	231	206	174	183	238	259	265	2.764
Tessalit	293	282	299	289	298	262	279	284	268	283	281	273	3.392
Tombouctou	283	288	297	283	306	284	294	281	279	291	287	276	3.447

Table 5: Monthly averages of daily global irradiation at five stations in Pakistan, measured in kWh/(m^2d) (Deutscher Wetterdienst, Hamburg)

Month	Peshawar	Lahore	Multan	Quetta	Karachi
January	3.00	2.52	3.51	4.01	4.54
February	3.98	3.18	4.31	4.60	5.20
March	4.95	4.97	5.41	5.54	5.35
April	6.07	5.68	6.24	6.74	6.32
May	7.13	6.30	6.71	7.95	6.63
June	7.28	6.17	6.56	8.27	6.53
July	6.50	5.56	6.39	7.44	5.61
August	5.84	5.28	6.12	7.08	5.16
September	5.37	5.10	5.81	6.85	5.78
October	4.49	4.33	4.83	5.83	5.50
November	3.58	3.39	3.99	4.71	4.80
December	2.99	3.03	3.30	3.87	4.27
Yearly average	5.10	4.68	5.27	6.07	5.47

Type of radiation: Solar cooking boxes make use of global (= direct and diffuse) solar radiation, while reflector cookers exploit practically only direct radiation (see Fig. 10). The difference between global and direct radiation depends on local climate, season, weather, etc. and can be considerable.

Location: Solar irradiation levels - and, hence, the usefulness of solar cookers - can vary substantially within a country; it all depends on the location of the project area, particularly if the country in question includes distinctly different topographies, as do Kenya, India and Mali, for example.

Cloud cover: The cooking process in a reflector cooker is interrupted when the sun disappears behind a cloud, and haze, which occurs in urban areas like Nairobi, makes the reflector cooker thermal output drop off considerably - even to the point of making it hard to keep food warm. During the rainy season, which lasts anywhere from 2 to 4 months in many countries, irradiation is so low that reflector cookers are of no use at all, and the usefulness of cooking boxes is limited, e.g. to keeping cooked food warm. The same applies to any cloudy day.

Time of day: Cooking with solar energy is limited to the "better" daylight hours, unless the cooker is a heat-accumulating type. Due to atmospheric attenuation, the solar radiation received during the first hour or so after sunrise and the last hour or so before sunset is too weak to power a solar cooker. Solar tracking helps little during the marginal hours. By contrast, heat-accumulating solar cookers are useful day and night.

Season of the year: In any area with pronounced seasonal variation in local climate and prevailing weather conditions, solar cookers can only be used in the summertime, but not during the winter or the rainy season.

Impact: Solar cooking is only possible in the presence of adequate insolation and good weather. In other words, it is decisively dependent on climate, season and momentary weather conditions. Conventional cooking facilities must be maintained nearly everywhere - and that constitutes a serious handicap for solar cookers.

3.3 Energy Demand

The purpose of solar cookers, of course, is to save energy in the face of a double energy crisis: the poor people's energy crisis is the increasing scarcity of firewood, and the nation's energy crisis is the growing pressure on its balance of payments. Solar cooker projects should be judged with that in mind.

Energy consumption: Compared to other nations, developing countries consume very little energy. For example, India's 1982 per capita energy consumption rate, at 7325 GJ, was one of the world's lowest. But the country's energy consumption rate is increasing nearly twice as fast as its gross national product. The same is true of most developing countries. In Pakistan, 60% of all energy consumed is commercial energy, and the remaining 40% is drawn from wood (26%), dung (8%) and agricultural residue (6%). According to reports from Mali, 93% of the energy consumed there is from noncommercial or semicommercial sources. In the Sudan, private households account for more than 3/4 of total energy consumption (78%), and the rest goes for transportation (11%), industrial production (6%) and agriculture (2%); the energy vehicles are wood (45%), charcoal (29%), other biomasses (9%) and petroleum (17%). In other words, the main energy vehicles are still the traditional ones. The fact that energy lost in the conversion of wood into charcoal accounts for about 1/3 of the country's total energy consumption is revealing. Such statistics usually exclude solar energy and "manpower" and are accordingly limited in their informational content.

Household energy: Private households account for 11% and 14% of commercial energy consumption in India and Pakistan, respectively. The statistical data vary widely from region to region and depending on the local climate. In Pakistan's lowlands, for example, the average annual per capita energy consumption rate comes to 2.9 GJ, rising to 15.1 GJ in the mountains. Most household energy derives from noncommercial sources, particularly from locally available renewable sources like firewood, scrubwood, brushwood, undergrowth, broken limbs and branches, dung from animals, agricultural residue like rice stalks and cobs, and spent tea and coffee bushes in Kenya. By comparison, grass reed and straw are used in Mali, corn stalks, sugar cane and cotton waste in Egypt, and sawdust in Pakistan. Different energy vehicles can have different uses. Agricultural residue, for example, is used in India as cattle fodder, but also as roofing material and, after composting, as a fertilizer. Whenever such sources of energy are used as fuel for cooking, they are naturally no longer available for use as, say, agricultural fertilizer. In India, more dung is burned in cooking fires than chemical fertilizer is used in farming. Thus, energy conservation via solar cookers would, at least in theory, have indirect positive effects on food production. Some 75% of all fuel burned in Pakistani households is from traditional renewable sources, above all firewood; and the firewood consists of more than just dead wood - green branches from trees and whole bushes are often included. The same is true of Kenya (cf. table 6). In the Sudan, more than 98% of all household energy comes from biomass. In the country's urban areas, charcoal is the predominant household energy vehicle. In appraising a solar cooker project, it is of decisive importance to know which source of household energy predominates: firewood, charcoal, petroleum, electricity, etc. - and what the prerequisites for and consequences of their use or nonuse may be.

Table 6: Energy consumption in Kenya according to sectors and energy vehicles (1980) /18/

	Firewood	Charcoal	Residue	Petroleum products	Elektricity	Total
	64%	6%	3%	25%	1%	100%
Urban households	2	58	--	4	19	6
Rural households	72	26	100	4	--	52
Agriculture	--	--	--	8	13	2
Industry	25	15	--	24	38	24
Trade	1	2	--	1	28	1
Transportation	--	--	--	58	2	14
	100	100	100	100	100	100

Energy consumption by the poor: The poor majority of the people in developing countries cover most of their energy requirement in a non-commercial (subsistence) way, using traditional, locally available sources of energy and their own physical labor. They simply cannot afford to buy any appreciable amounts of commercial energy. The use of firewood is often identical with poverty; depending on the country in question, the middle classes are more likely to use charcoal, while the upper classes have access to gas, electricity and kerosene (cf. table 8). Moreover, energy consumption out in the country and among the poor is characterized by lower efficiency than in town. It is interesting to note that biomass consumption in rural Pakistan does not drop off as family income increases, but instead increases (or, in towns, remains the same). The logical consequence is a relative shortage of fuel for use by the poor, whose living conditions deteriorate even more as a result. Solar cooker projects could at least try to compensate.

Procuring energy for cooking: As long as there is an ample supply of dead wood in the near vicinity, solar cookers will have little chance of acceptance, especially since the task of collecting firewood often has a social function in that it provides an opportunity for communication. Depending on the indigenous system of values, gathering firewood is seen as a job for women, e.g. in Mali, or for men, e.g. among Afghan refugees in Pakistan, or even for the older children, e.g. in Pakistan. In some Islamic societies, men are responsible for gathering wood, while in others - like in Central Sudan - 61% of the women, 57% of the children and 48% of the men participate as long as wood is available within a reasonable distance; otherwise, charcoal is given preference. In India, it is a question of caste and region as to whether women or men should gather the wood: in the lower castes, it is the women who are mostly responsible for gathering wood. Such arrangements are not without impact on the social acceptance of alternative sources of energy, e.g. solar energy, in cases where traditional fuels have become scarce, thus costing more time to collect them or more money to buy them. Another factor that should not be overlooked is that many of the poorest people gather and sell wood as their only source of income - and solar cooking projects are perceived as a relevant threat.

Table 7: Energy consumption indicators for the Sudan /109/

	Eastern Sudan		Southern Sudan		Gezira
	rural	urban	rural	urban	rural
	%	%	%	%	%
Energy vehicle					
firewood	97	81	100	76	65
charcoal	73	96	11	82	92
agric. residue	43	--	--	35	86
kerosene	32	73	--	68	12
diesel	66	35	33	27	8
electricity	1	42	--	23	8
dung	15	--	--	--	47
vegetable oil	--	--	--	--	41
Procurement of wood					
buying	58	69	--	52	18
gathering	38	12	100	18	33
Energy consumption (kg/d per capita)					
wood	1,0	0,64	3,6	1,2	0,86
charcoal	0,45	0,36	0,18	0,32	0,45
Wood gatherers					
men	63	100	11	17	48
women	28	100	89	67	61
children	35	100	--	17	57
Time spent gathering wood					
short	40	100	56	67	57
less than 1/2 day	28	--	44	33	17
1/2 day	33	--	--	--	4
more than 1/2 day	--	--	--	--	17
Wood needed for					
cooking	77	14	100	68	78
other uses	23	86	--	32	22
Procurement of charcoal					
buying	73	96	11	82	88
producing	--	--	--	--	4
not used	27	4	89	18	8
Charcoal needed for					
cooking	88	100	--	98	87
brewing tea and coffee	95	100	100	96	84
ironing	79	100	100	10	91
other uses	4	12	--	17	9
Agricultural residue needed for					
cooking	13	--	--	--	64
other uses	29	--	--	35	--
not used	58	100	100	65	36
Kerosene needed for					
cooking	8	42	--	13	--
illumination	100	100	--	100	73
lighting fires	58	42	--	16	27
Electricity needed for					
cooking	--	--	--	13	--
illumination	100	100	--	73	100
other uses	--	--	--	17	--

Table 8: Household energy consumption in Kenya for different income classes (1980) /18/

Group	Income (KSh p.a.)	Households (millions)	Firewood	Charcoal	Kerosene	Elektricity
					(GJ x 10^6)	
1	0 - 3,100	.09	1.7	0.9	0.0	0.0
2	3,100 - 9,100	.16	1.3	3.2	0.9	0.0
3	9,100 - 18,200	.17	1.0	4.5	1.0	0.0
4	18,200 - 54,600	.11	0.3	2.8	0.6	0.3
5	54,600 -	.05	0.0	0.7	0.5	0.6
Totals		.58	4.3	12.1	3.0	0.9

Cooking-energy quantities: The daily fuel requirement varies according to the kind of food being cooked and the number of warm meals. In the typical LLDC, each native burns one ton of firewood each year.

- In India, the average family needs somewhere between 3 and 7 kg of wood per day; in the cooler regions, the daily firewood demand varies between just under 20 kg in the winter and 14 kg in the summer.
- The nomads in northern Kenya use 70% less firewood during the rainy season, when milk is their staple food item, as compared to maximum wood consumption during the dry season, when mostly meat is eaten.
- In the southern part of Mali, the main dishes (corn, rice, millet) and the way they are prepared make for a higher level of firewood consumption than in the northern part of the country, where fish and roasted meat/fish are the staple items. The average 15-member (!) family burns about 15 kg of wood each day.
- A survey conducted in an Afghan refugee camp in Pakistan showed a daily firewood demand of 7 kg per family, but a different survey put the demand at 9...10 kg per family and day. More than half of the wood used in the average household goes for baking, and the remainder is used for cooking. Additional wood is needed for heating in the wintertime, of course.

- A family of five in the Sudan consumes roughly 2.5...3 kg of charcoal every day. According to the World Bank, the annual per capita firewood consumption rate ranges from 1.2 kg in towns to 2 kg per day in rural areas.
- Conclusions: The above examples indicate quite clearly that the required amounts of cooking energy are extremely variable; only painstaking local-scale investigations are able to pinpoint how much cooking energy can be saved by using solar cookers.

Using cooking energy: The example for Pakistan shows that differentiation should sometimes be made between fuel used for cooking - of special relevance to solar cooking - and fuel used for baking. In many places, namely, more energy is consumed for baking than for cooking. In the central region of the Sudan, 78% of all families used wood for cooking, and 87% (also) use charcoal.

Time expenditure for procuring cooking energy: The women in Mali reportedly spend about three hours every day gathering wood. Reports from Pakistan mention up to 15 hours per week, and between 10 and 20 hours per week are cited for an Afghan refugee camp in Pakistan. Rural families in India have to invest about six hours each day (!) for gathering wood. In the Sudan, the amount of time needed for gathering wood is steadily increasing, so that wood is gradually becoming something of a nondurable consumer item for poor people, too, which puts it in competition with charcoal, the latter being easier to haul. The potential time savings offered by solar cookers appear substantial.

Cost of cooking energy: Table 9 shows the retail cost of various cooking energy vehicles in certain African countries. The prices are seen to differ widely, depending on the country, the prevailing energy policy and the individual region. It is also a question of which energy vehicles are subsidized to which extent, e.g. kerosene, which enjoys a 30% subsidy in Pakistan's Orangi Project. Here, the average expenditure for fuel comes to about 25% of a family's monthly income. In Bamako, Mali, the average family spends between 15% and 20% of its earnings on fuel, i.e. wood and charcoal. In Pakistan, the cost of firewood has stayed in step with the consumer price index (c.p.i.) during the 1970s, but has pulled ahead considerably during the '80s. In the Sudan, firewood and kerosene prices stayed well below the c.p.i. between 1975 and 1984, while the cost of firewood has increased sevenfold (with wild gyrations) in the course of the past 8 years. Many families in Khartoum spend anywhere from 20% to 30% of their household income on energy. If cooking energy keeps getting more scarce and accordingly more expensive in India, the traditional noncommercial sources of energy will soon be more expensive than food. Anyone in the situation of having to buy cooking energy stands to profit solar cookers.

Table 9: Retail cost of various energy vehicles for household cooking in urban areas of several African countries /51/

	Ethiopia 1)	Burkina Faso 2)	The Gambia 3)	Kenya 4)	The Niger 5)	Senegal 6)	Uganda 7)
Wood ($/kg)	0,077	0,095	0,137	0,018	0,058	0,061	0,034
Charcoal ($/kg)	0,545	0,237	---	0,092	0,145	0,152	0,109
Kerosene ($/l)	0,564	0,488	0,544	0,348	0,401	0,424	0,400
LPG ($/kg)	0,603	1,007	1,048	0,627	0,954	0,092	0,700
Elektricity ($/kWh)	0,072	0,261	0,193	0,064	0,170	0,240	0,006

1) UNDP/World Bank, Ethiopia; Issues and Options in the Energy Sector; May 1984

2) Document de l'Institut Voltaique de l'Energie; 1983. Lepeleire et al.; Project CILSS - Foyers Ameliores - Elements d'Evaluation - Suggestions; 1984

3) UNDP/World Bank; The Gambia: Issues and Options in the Energy Sector; November 1983

4) IPC; Charcoal Production and Research Activities within the Special Energy Programme Kenya; June 1984

5) UNDP/World Bank; Niger: Issues and Options in the Energy Sector; May 1984

6) UNDP/World Bank; Senegal: Issues and Options in the Energy Sector; July 1983

7) UNDP/World Bank; Uganda: Issues and Options in the Energy Sector; July 1983

Firewood: The prime function of solar cookers is to help reduce firewood consumption, since most cooking fires are still fueled with firewood. The trouble is, firewood is usually quite inexpensive in comparison with kerosene, bottled gas or electricity (based on relative energy content). Increasing, uncontrolled felling of wood for people's own use and for selling are a main cause of deforestation, desertification, erosion, receding groundwater levels, ... - in short: it has long-term adverse effects on the ecological balance. Pakistan's meager forest heritage and rampant deforestation in Kenya show that such fears are well-founded. If denudation of the Sudan's forests continues at the present rate, they will be gone by the year 2005. As of this writing, 70% more wood than allowed was being felled. Massive diffusion of solar cookers could have positive effects.

Energy policies: National energy policies in Third World countries with little or no oil resources are usually based on the substitution of oil and/or natural gas, possibly in combination with hydropower, because expensive energy imports have a devastating effect on such countries' trade balances. The quest for ways to save energy by using it more efficiently and by developing alternative sources of energy, e.g. solar energy, or by boosting the available manpower (the main source of energy in poor societies) through more and better food, health care and education is given less priority, even if it is mentioned in some development plans, in which case the plans may be regarded as exemplary from the development-political standpoint. On the whole, solar cooker projects could, at best, contribute little toward a national energy policy. But they could make a very substantial contribution toward improving the living conditions of the poor and helping them overcome their own energy crisis - if they were accepted.

3.4 Social Situation

Decisions for or against solar cookers are predisposed by the social situation, living conditions and customs of the native populace.

Poverty: If the "poor" majority of the Third World's people is the target group, then solar cooker projects must be first and foremost to the benefit of the rural population. In India, Kenya, Mali and the Sudan, about four out of five people live in rural areas - usually with more than half of them at or below the poverty line - as farm workers, tenant farmers, subsistence farmers, craftsmen, (itinerant) traders, often enough within the subsistence economy. More than 80% of all arable land in Mali and the Sudan is used for subsistence farming. The low, unstable - and in the Sudan even sinking - purchasing power of the rural population is an important boundary condition for the introduction of solar cookers.

Population groups: The conditions for introducing solar cookers may be more advantageous for some population groups than for others. Urban middle-class families with modest savings, for example, are more likely to have an open attitude toward experiments. On the other hand, solar-cooker dissemination efforts require the aid of certain groups, e.g. women's groups, artisans' groups or other groups in influential positions, like ethnic leaders (maliks) and religious authorities (mullahs) in Pakistan. For a solar cooker project to reach its target group, it can hardly do without such intermediary target groups, who then proceed to pass the idea on to the intended users. However, it is not always easy to choose the "right" intermediary target group - consider, for example, the caste system in India, where the wrong choice could do more harm than good to a dissemination program.

Systems of values: When a patriarchal social order allows only men to make important decisions, or when religious norms like the observance of Ramadan call for strict fasting from dawn to dusk, this can have substantial effects on the usefulness of solar cookers. In polygamous Mali households (about 30%), each wife has her own hearth and "pantry" within the estate, even though they all take turns cooking. In Kenya, too, each wife in a polygamous household has her own separate hearth. In the Sudan, the adult women in large rural families take turns cooking. Such cultural and socio-structural aspects are of great significance in connection with the introduction of solar cookers.

<u>The woman's role:</u> Despite similar systems of values, the woman's role can differ decidedly from one region to another within a given country.

- <u>Mali:</u> About 65% of the Mali population are Muslims; despite the Islamic influence, however, women have managed to maintain some degree of economic independence. Since men and women have separate spheres of everyday life and work, each has his/her own separate responsibilities and source of income. While the men derive most of their income from selling agricultural produce, the women are responsible for growing/procuring the family's food, with the notable exception of grain. Basically, the actual economic and social role of Mali women depends on their ethnic affiliation and main source of income, e.g. from farming or stock breeding.
- <u>Kenya:</u> Kenyan women also play a leading role in their country's development process - probably thanks to their gardening/farming activities and related income.
- <u>Pakistan:</u> Women's activities in Pakistan are confined mostly to the household. Also in the Afghan refugee camps, Islam exerts a substantial influence on everyday life and on the social order. Social life is strictly patriarchal and completely excludes women from all public activity. It would be just as impossible for them to participate in a public solar cookers demonstration or other event as it would be for them to engage in household activities outside of the courtyard area: fetching water or wood, shopping for groceries, etc.
- <u>The Sudan:</u> Sinking real income in the Sudan has led to closer integration of women into working life.
- <u>Conclusions:</u> A woman's role - especially her voice in family affairs, use of income, etc. - has a substantial effect on the introduction and use of solar cookers; consequently, the actual effects of dissemination campaigns and related educational programs cannot help but vary accordingly.

3.5 Dietary Patterns

Solar cooker projects can hardly be expected to change people's dietary patterns, but they must at least be suitable for use in cooking most of the main dishes.

Staple foods: The staple foods of a population can differ quite substantially, depending on the country, region, local traditions, crop harvests, availability, season of the years, etc. No solar cooker project should be implemented without a good working knowledge of the prevailing staple foods. Five examples:

- India: Rice and wheat are the two staple foods. Most rice is cultivated in the southern part of the country, and most of the wheat is grown in the North. Legumes, chick peas (grams) and various vegetables are also eaten. Meat plays a subordinate role. In the poorer households, millet (panic grass) is eaten in place of rice and wheat. Eggs, milk and other cereal products are eaten more in well-to-do households than among the poor.
- Kenya: The principal food in Kenya is corn (maize), followed by beans, various leafy vegetables and tubers. In some areas, goat meat and poulty, potatoes and eggs, cooking bananas and various vegetables are eaten when available. The Nomads in northern Kenya live on milk, meat, corn meal and little else. They eat mostly meat during the dry season and milk (products) during and for a while after the rainy season. Due to lack of infrastructure, corn is not always available. Corn consumption is concentrated mostly in the interseasonal months when lactation has not yet begun, but the consumption of meat already has to be reduced.
- Mali: Millet, corn and rice are the principal foods in Mali. To the extent available, and depending on their ethnic background, people also eat vegetables, meat, fish and cabbage, the latter seldom. In southern Mali, millet and rice are preferred, along with peanuts, ocra, sweet potatoes and fish. The leaves of the fiber plant hibiscus cannabis are widely used in preparing sauces. A fat traditionally extracted from the fruit of the shea butter tree is used as cooking oil. In the central area, squash, tomatoes, sweet

potatoes, rice and onions are very popular. The Nomads in northern Mali live chiefly on milk products, meat and fish.

- Pakistan: In this country, the two staple foods are rice and wheat. Lentils, beans, various vegetables, herbs, and meat are prepared together in stew fashion. While fish is an important source of nutrition for people living in Karachi and other coastal areas - particularly on the "meatless" days prescribed by the government - many people in the mountainous regions live almost exclusively on fruit, vegetables and grain. Families with above-average incomes also consume eggs, salad, fruit and wheat products.
  Afghan refugees in northern Pakistan have bread, vegetables, meat, and tea as their primary foods /184/.
- The Sudan: The staple food is durra (primarily sorghum plus some millet), in addition to tomatoes, broad beans, potatoes, lentils and vegetables, usually in the form of stew, which may also contain mutton and goat's meat. Due to sinking income, a bean dish called foul is gradually disappearing from poor people's menus.
- Conclusions: People have to be able to cook their staple foods in solar cookers. As long as the staple foods are rice, cereal, grain, noodles, vegetables, potatoes and mixed dishes, e.g. stews, solar cookers can serve well, though care must be taken to differentiate between the various types. A low-temperature cooking box, for example, may be just right for social groups whose main foods are rice and vegetables, but it would not be of much help when the cook needs temperatures hot enough to simmer cooking oil or to bake crispy bread. The utility value of a particular solar cooker is more or less limited by the relative importance of cooking, boiling, steaming, roasting, baking or grilling.

Meals: The most important thing is that the main dishes with the staple foods can be prepared using solar cookers. To judge this properly, one must give due consideration to the people's morning, noon and evening eating habits, i.e. to the time of day when the main meal is taken:

- Breakfast: In rural communities, breakfast may well be more important than the midday meal (lunch). In India, to the extent that breakfast is not eaten cold - and this is where solar cookers come in - it may consist of chapatis (a pancake-shaped unleavened bread, usually made of wheat flour and baked on a griddle, and which is common in northern India) milk or milk products, and sometimes tea. A typical breakfast in Mali consists of millet gruel with milk and sugar and salt. Bread (naan) left over from the day before with tea and milk is the basic breakfast in Afghan refugee camps in Pakistan. Urban families in Pakistan often have fried eggs for breakfast. Pakistanis of Indian descent often supplement their breakfast tea and chapatis with a boiled stew consisting of vegetables, potatoes and meat. In Kenya, breakfast consists of tea with milk and sugar and a thin mush cooked in milk or water. As in all other countries, there are important differences between various social groups: The Nomads of northern Kenya eat a soup of milk, corn meal and sugar (uji) and drink tea with milk. In the Sudan, a typical breakfast consists of millet gruel, vegetables or warmed-up foul, a fresh batch of which is prepared twice a week. Tea with milk is normally drunk for breakfast. Bread is sometimes eaten as a side dish. Well-to-do families add tomatoes, vegetables, eggs, cheese or liver, while poor families make do with millet, milk and an occasional piece of jerky. The useful effect of solar cookers depends a lot on the importance of fresh-baked bread at breakfast time.

- Lunch/supper: The noonday meal often differs only insignificantly from the evening meal. That being the case, the expediency of using a solar cooker when the sun is at its highest has both advantages and disadvantages. Many poor people eat little or nothing for lunch, saving what they have for supper. In Pakistan, for example, most people eat a modest lunch consisting of a vegetable stew, meat with chapatis and/or rice. Pakistani families of Indian descent often content themselves with a snack of breakfast leftovers. Some farmers and farm workers take "lunch boxes" to the fields. Frequently, though, farm hands buy their lunch from the landowner - at exorbitant prices. Mulach, a stew consisting of meat, onions, vegetables, garlic and oil, is often eaten for lunch in the Sudan; kisra, a kind of sorghum bread, is eaten with the stew. Many poor people have only tea and milk for supper, but others eat the same dishes as for lunch. Such diverse conditions naturally have an impact on the use of solar cookers.

- Main meal: Knowledge of what is required to prepare the main meal - or the "national dishes" - is of special importance, since solar cookers will be expected to have what it takes in that connection. If the dish in question is boiled, e.g. foul, a bean dish popular in the Sudan, or dal, an Indian dish based on legumes, then solar cookers are just what is needed; the same applies to kitheri, a stew consisting of beans, tomatoes, cooking bananas and squash that is popular in Kenya, and to khichadi, an India stew made up of rice and legumes. Obviously, it would be impossible to assess the acceptance of solar cookers without knowing how the staple dishes are cooked.

Cooking staple dishes: The preparation of main dishes differs according to local tradition and the type of cooking facilities used. Additionally, the cooking temperatures and manner of cooking (boiling, roasting, baking, grilling) also differ from case to case. Some examples:

- Mali: Tot is prepared by putting a thick flummery in roughly three liters of boiling water and stirring steadily to obtain a firm mass. The stiffness can be increased by adding powdered baobab leaves. Couscous, by comparison, is a more complicated dish to prepare: a coarse-grained, dry noodle substance made of corn or durum wheat (semulina) is put in a perforated bowl and allowed to swell and cook in the low steam heat over a pot of simmering water. Once in a while, it has to be loosened up by hand. In some places, the semulina is cooked directly in boiling water until finished. Rice is prepared like couscous or in boiling water. The sauces are made by searing vegetables and meat in hot cooking oil, quenching with water and then cooking until finished. It takes about 60 minutes to cook tot and sauce and between 80 and 90 minuntes to prepare couscous. In northern Mali, meat and fish dishes are roasted, fried or smoked.
- The Sudan: The bean dish foul is easy to make, because the beans need only be cooked slowly without stirring. Mulach, by contrast, is more complicated in that the onions are first

seared in hot oil, after which meat, salt and water are added and left to boil; after a while, vegetables, tomatoes or alimentary paste are added, after which frequent stirring is necessary. Aseeda, a millet pudding, also has to be stirred frequently. Sudanese "pancakes" (kisra), made of fermented millet powder, are usually cooked on a griddle-like sheet of metal.

- <u>Kenya:</u> Ugali is prepared by placing coarsely ground corn meal in boiling water and stirring vigorously for a few minutes. By comparison, it takes up to three hours to cook kitheri stew. The ingredients for a meal are frequently cooked in succession, with the finished ones kept warm by placing them on top of the hot pot.
- <u>India:</u> The unleavened bread chapati is baked on a griddle or in a skillet over a hot fire for several minutes and then roasted briefly in the open flame.
- <u>Pakistan:</u> The stew (vegetables + meat), as well as the rice, is stirred steadily. Indian families use a lot of fat for cooking. All ingredients (except the rice) are first fried briefly in cooking fat and then boiled in water. Baratta is prepared on a very thin sheet of metal called a tawa. The dough is baked ahead of time and then fried in lots of fat. Chapatis are either also prepared on a tawa or baked in a tandoor (a built-in adobe oven). The dough is slapped onto the inside of the very hot oven from the top and turned after a minute or two. Some families in one compound have communal tandoors in which several families can bake their chapatis together. In the larger towns, however, people often buy their chapatis in bakeries.
- <u>China:</u> The staple foods require boiling, steaming, roasting, etc. for preparation. This means that high temperatures and high thermal power are necessary. These requirements are easily met by reflector cookers, while in this case cooking boxes do not perform satisfactorily.
- <u>Conclusions:</u> The above examples clearly show how solar cookers have to "adapt" to how the main dishes are traditionally prepared. Consequently, most people tend to use a solar cooker for just a few of the more appropriate dishes. Indian ATRC cooking boxes, for example, are used mostly for cooking rice, legumes and vegetables; in the relevant

project, among the users, 20% make use of their solar cooker only for certain dishes, and 60% use it only for one meal a day.

Bread: Most solar cookers are of little use in areas where baked bread is the main staple. While solar cooking boxes will bake bread, at least in principle, they cannot produce a crispy crust. Afghan refugees in Pakistan, for example, eat large amounts of nan (unleavened bread made of wheat flour) practically for every meal. According to a 1983 survey of 498 households, the average family consumes 20 loaves of unleavened bread per day. Half of the women interviewed said they bake twice a day, and one in five bakes a full day's supply at once. Most nan is baked in traditional tandoors. Past attempts to replace wood with other kinds of oven fuel, e.g. briquettes or rice hulls, failed due to slight discoloration of the bread. Unleavened bread has to be baked at a very high temperature and requires heat distribution on relatively large surfaces; the cost of the technical necessities would very probably surpass the low solar-cooker price limit for poor social groups. On the other hand, some baked goods turn out especially good in solar cookers, as substantiated by reports from the Sudan.

Side dishes: In some countries, Mali for instance, the usefulness of solar cookers seems to be restricted to the preparation of entremets and sauces/gravies, as opposed to main dishes. That is surely an impediment to their acceptance. Solar cookers really should be well-suited - or even better-suited than traditional means - for cooking at least certain kinds of food, or people won't try them in the first place.

Taste: Indian housewives contend that unleavened bread loses its typical flavor if baked in a solar cooker. Indeed, they say, other dishes also come out tasting different if they are cooked in, say, an ATRC cooker. Under such circumstances, solar cookers can hardly be expected to find acceptance. The extent to which the taste of Sudanese food is altered by cooking in a solar cooker is a point of controversy.

Beverages: At first glance, solar cookers appear to be just the thing for preparing beverages, above all tea. In some cultures, however, the men like to spend lots of time preparing and drinking tea around an open fire - in the shade. In Mali, where the practice has taken on a nearly ritual character, this has had a definite limiting effect on the use of solar cookers. The same is true in the Sudan, where the low efficiency of the most commonly used type of solar cooker makes it unsuitable for brewing tea quickly - and tea is taken with all meals as well as at dawn and dusk.

3.6 Eating Habits

In evaluating the usefulness of solar cookers, one should keep in mind that not only what people eat is important, but also how and when they eat it.

Mealtimes: Just when meals are cooked and eaten is more or less inflexibly dependent on the family's workaday routine.

- Breakfast: Since most of the rural population is accustomed to working "from sunup to sundown"", they eat breakfast at about dawn, i.e. around 6:00 or 7:00 a.m., possibly as late as 8:00 a.m. The earlier the breakfast, the less useful the solar cooker, since solar radiation is still rather weak early in the morning. This can be especially problematic during Ramadan, when breakfast has to be eaten before the sun comes up. Most families in India forego an early breakfast in favor of brunch, thus saving the midday meal. Late breakfast between 9:00 and 10:00 a.m. is also customary in the Sudan. Due attention should be paid to such particularities, which can greatly enhance the utility value of solar cookers for breakfast.
- Lunch: As a rule, lunch is eaten sometime between 1:00 and 2:00 p.m., sometimes as early as noon. Here, too, there are some major exceptions to the rule, depending on the social group in question and their daily routine. In some parts of India, people eat lunch (brunch) around 10...11 o'clock in the morning, while the Sudanese tend more toward 2...4 in the afternoon. Afghan refugees in Northern Pakistan prepare the

main meal of the day between noon and 1 o'clock. But for evaluating the usefulness of solar cookers, breakfast and supper are more important than lunch with regard to timing.

- Supper: In practically all developing countries, supper-time has to wait until the workers have come home from the fields. In India, Mali and the Sudan, supper is often served as late as 9:00 in the evening.
- Conclusions: The traditional mealtimes cannot be altered for the sake of solar cooking, because they are firmly embedded in long-standing rhythms of living and working. It is therefore not surprising that 60% of those participating in one Indian solar cooker project cooked only a single "solar" meal per day. If, however, the solar cooker is a heat-accumulating type that collects solar energy independently of when the heat will be needed, then it doesn't much matter when meals are eaten.

Common meals: Since different members of the family have different work to do - housework, fieldwork, paid labor, going to market, attending school, ... - supper is often the only common meal of the day. Communal meals - with or without neighbors - are only possible on weekends and holidays in some places, like in Kenya where a shortage of arable land forces many men to take jobs in town or on distant plantations, while the women stay home and tend the fields. Lunch is often warmed up for the children when they come home from school in the late afternoon. All in all, the number of dishes prepared in the course of a day can vary widely, depending on the momentary situation. In Kenya, most meals are cooked for 6 to 10 people; in Pakistan and the Sudan between 10 and 15 people eat together in rural areas, but only 6 to 8 in town; Indian women cook for anywhere from 5 to 15 people. Such guideline values can fluctuate considerably, depending on the ethnic group, religion, time of day and day of week. If at all possible, a solar cooker should have enough capacity to serve a good number of people at once, not just the average family.

Eating sequence: In some strictly Islamic societies, e.g. in the Sudan, the man, as head of the family, has the privilege of eating first - served by his wives, daughters and daughters-in-law; then come pregnant and nursing women and children, then the other women, and finally the adolescents. In case of famine, this can lead to undernourishment, particularly among female adolescents and women in general. In some places, the men are served by women, while in others the men will only accept male servants and the women eat somewhere else, i.e. not in the man's presence. The fact that meals often have to be cooked "in shifts" instead of for the whole family at once must be given due consideration in connection with solar cookers.

3.7 Cooking Habits

How the food is cooked, i.e. how heat is applied via a liquid (water, fat) or air (baking or grilling), influences the appropriateness of solar cookers in the Third World. For women, who are nearly always responsible for cooking, often assisted - but rarely displaced - by their eldest daughters, the value of a solar cooker depends on how well it accommodates conventional cooking habits while saving a considerable amount of work.

Accessibility: Many dishes require frequent stirring. It must be easy to add more ingredients while the food is cooking. Tasting/seasoning must also be possible, of course. With some types of solar cookers, though, it is not possible to get at the food without having it cool off too much. For example, it is usually not possible to stir food in a box-type solar cooker - the pot has to be taken out of the cooker for stirring. That naturally has an adverse effect on acceptance, despite the time-saving advantage of cooking boxes, namely that the food can be left to cook by itself with no stirring necessary. In fact, those advantages are not always realized. According to reports from Pakistan, it was very difficult to convince women of the fact that the food in a solar cooking box need not be stirred continuously. The women involved in Pakistan's Orangi Project criticized the fact that the food cannot be taken out of the box until it is finished - mainly because the customary main dishes, if prepared according to conventional methods, require several successive cooking operations.

Cooking time: It takes up to three hours to prepare dishes like kitheri and mulach (staple foods in Kenya and the Sudan, respectively). When different members of the family eat at different times, the stove may be in service anywhere from 6 to 8 hours a day. In the event of delays due to moderate temperatures caused by low thermal output, high heat losses, or low irradiance (glass cover, dusty reflectors, cloudy day), cooking can take much longer than usual. That naturally retards acceptance, e.g. in the case of the Sudanese solar cooking box. In other words, fast cooking is a very important acceptance criterion.

Interruptions: Some solar cookers make it very difficult, if not impossible, to interrupt cooking in the course of complicated meals, because it is hard to get the pots out of the cooker, or thermal losses would be too high. That, too, has a bearing on acceptance.

Attendance: With some types of equipment, e.g. solar cooking boxes, the cook cannot tell when the food is finished. Opening the box would waste a lot of heat - and time. But when clouds, haze and shadows retard the cooking process more or less erratically, this can be a real problem. On the other hand, solar cooking boxes do not require attendance. The advantage is that the cook can go do something else like work the fields, tend the garden, prepare other kinds of food, wash the dishes, nurse a baby, ... But that only comes to bear when women are actually able to take advantage of it, i.e. when cooking the meal does not involve too many intermeshed processes.

Cooking processes: Overlapping cooking processes are necessary for many staple dishes. In Pakistan and the Sudan, some kinds of food are first fried and then cooked under constant stirring. When using a solar cooking box, this is impossible, but it works with a reflector cooker. Conversely, of course, food that can be cooked in a single step e.g. without having to be fried first, can easily be prepared in cooking boxes. Meals that are cooked in several pots at once instead of in succession are also good in that sense.

Size of morsels: For the acceptance of solar cookers, food that has been cut up into small pieces is better than food that is cooked in large chunks.

Warming: Some dishes have to be kept warm for a certain length of time, either until the other dishes are ready (a cooking box will do the job) or until evening (in which case a heat-accumulating solar cooker or a non-solar insulated "hot box" may be necessary). Reflector cookers can do neither.

Burning: One of the foremost advantages of some solar cookers, especially of box-type solar cookers, is that they won't overcook or burn the food. On the other hand, that advantage is gained at the cost of lower temperatures and longer cooking times.

Baking: Some devices, like the Tube Solar Oven, are equally well suited for cooking and baking. Depending on the prevailing dietary patterns, that can be of special advantage as long as solar baking has no adverse effect on taste and when the customary food preparation processes involve both cooking and baking. In solar cooking boxes, baking is possible, too. But the low temperatures don't allow the generation of a crispy crust.

3.8 Use of Stoves, Fireplaces and Hearths

Over the centuries, stoves have become accommodated to the eating and cooking habits of their users - and vice versa. For solar cookers, there is no getting around a comparison with traditional stoves.

Types of stoves: The most common type of "stove" is an open fire, sometimes supplemented by other cooking facilites. Some examples:

- Kenya: Kenyan stoves are appropriate to the climatic, economic and cultural situation. Most women do their cooking over a three-stone fire like the one shown in figure 16. Those who can afford one use a little round wood/charcoal-fueled sheetmetal stove called a jiko. Some Nomads build little baked-clay (adobe) walls around an open fire, and others cook in a kind of adobe pit. For Nomads and for market women, mobility is a prime criterion for a hearth/stove. In areas with pronounced seasonal changes in temperature, mobile stoves are also prefered, since they enable cooking indoors or outdoors, as desired. In cool weather, the hearth also serves as a heater. Many families also use it as a source of light, making it the centerpoint of social life.
- Pakistan: The cooking facilities used by Pakistani families also differ according to region and family income. Most rural housewives cook their meals on one or more chimneyless adobe stoves. In families with adequate income, the women cook on a gas- or kerosene-fueled metal stove. Practically every rural Pakistani family has its own oven for baking, in contrast to Indian families, who frequently bake their bread on top of the adobe stove. The oven also serves to keep the food or tea water warm for extended periods. At higher elevations and in other relatively cool areas, the hearth also serves as a source of heat, but rarely as a source of light or as a family meeting place. Afghan refugees in Pakistan cook on an open three-stone hearth (indoors in winter and outdoors in summer) or on a self-constructed horseshoe-shaped adobe stove. Some families also have a metal stove or kerosene cooker. Most refugees are given a free kerosene cooker. Unfortunately, the cookers in question are of poor quality,

smoky and bad for the taste of the food. Consequently, some families hardly use their cooker at all, or even buy a better one at the market. Women use the cookers mainly for boiling water for tea and for preparing light dishes, e.g. vegetables.

- India: A small chimneyless earthen or brick stove called a chulha is the customary cooking device among Indian women. A given family can have either several earthen stoves for one pot each or one big chulha for several pots at once. Small, charcoal-fueled metal stoves that double as space heaters are also used in cooler regions.
- Mali: Most Malian women cook on three-stone hearths. Additionally, there are various types of earthen stoves for one or more pots, plus various metal/cast-iron cookers. The women often use several cookers at once if necessary. Nearly all Malian hearths are portable and can therefore be used indoors or outdoors, depending on the weather and lighting conditions. In hot areas, the women prefer to cook indoors to avoid the high outdoor temperatures. In cooler areas, the hearth also serves as a space heater. In the evening hours, the open fire becomes an important source of light and a meeting place for family members and neighbours.
- The Sudan: Three-stone hearths are customary in rural areas. They are often accompanied by tin stoves for charcoal. Also customary are baking tins over a wood fire (cf. Fig. 17). As a rule, a family has several different lightweight hearths.

Side effects: The main lesson to be drawn from the above examples is that traditional hearths can be quite versatile and have some positive side effects, e.g.:

- versatility: traditional hearths are more than just means of cooking food; they also provide heat and light;
- side effects: preservation of food by smoke from an open fire increases storability and, hence, makes for a safer supply of food; the smoke also drives away insects;
- communication center: in many societies, hearths have an important social function in that they serve as centers of communication, especially in the evening;

- conclusions: solar cookers have to compete with those merits. The extent to which they can or cannot compete depends on criteria like the following:

Figure 16: Three-stone hearth

Figure 17: Charcoal cooker and baking tin in the Sudan

<u>Size of hearth:</u> Sometimes, setting up and operating a solar cooker takes a considerable amount of (extra) space that is not always available, especially in densely populated urban quarters. Some solar cookers - especially reflector cookers - are quite voluminous and require an accordingly large area.

<u>Setting up:</u> While some solar cookers are installed for stationary operation, others have to be set up anew every morning. Many Indian families that own an ATRC cooker, for example, complain that it is too heavy to be carried in and out of the house every time it is needed for cooking. The solar cooker project that was carried out among Afghan refugees showed that many would simply use their traditional hearths if no one was around to help carry out the box in the morning. Many devices are just plain too heavy for women. Others, like the one used in the North Horr, Kenya Project - are too bulky for easy transportation. Mobility is particularly important for some social groups: polygamous women, market women, field workers. Accordingly, solar cookers must satisfy quite stringent demands with regard to mobility.

<u>Stability:</u> Sometimes, mobility conflicts with the stability that is needed for, say, stirring the food, keeping the cooker from being knocked over by children or domestic animals, etc.

<u>Cooking height:</u> Coming home from a long, hard day out in the fields, women in the Third World prefer to sit down or squat for cooking. Some solar cookers, however, can only be operated standing up; in some cookers, the pot is suspended at shoulder level. That, too, impairs their social acceptance.

<u>Location:</u> Women customarily do their indoors cooking in a walled-off cooking area or in a separate room or hut. For some Nomads in northern Kenya, it is strictly taboo to cook or eat outdoors. According to one adage, "only hyenas eat outdoors". But it lies in the nature of (most) solar cookers that they be used outdoors, no matter how disdainful that may be in Kenya. It is also frowned upon in India, not only because a stove standing outside is easier to steal, but also because proper

weather protection is hard to come by: the stationary IGDP-type cooker (Dhauladhar solar cooker) in India has to be covered with a tarpaulin and carefully serviced during the rainy season. Very few solar cookers, e.g. the cooking part of the Solar Hot Plate Cooker, can be used indoors.

Cooking in the shade: As a rule, the solar cooker has to stand out in the sunshine, while the woman doing the cooking normally would prefer to be either indoors or at least outside in the shade of a tree, where there is also room for chatting with other women and the children. A reflector cooker, however, is merciless: the woman is forced to stand out in the sun, because the cooker has to be kept properly adjusted, and the food has to be stirred periodically. Only few solar-cooking devices will allow the work in the shade, e.g. the heat-accumulating ISE solar cooker, the Solar Hot Plate Cooker, the Convective Solar Cooker, etc.

Cooking capacity: The question of how well the amount of food that can be cooked at once in a solar cooker, will meet the requirements of an average target-group family is an important criterion for acceptance. The aforementioned eating habits play an important role with respect to individual family members and social groups. Solar cookers in North Horr, Kenya, are too small for most rural families, because their cooking capacity is limited to three kilograms. Large families in India have the same problem with ATRC cookers. In many projects, the capacity of the cooker seems to be oriented more along the lines of a four-member urban family than for a typically large (10-15 people) rural family.

Loading capacity: Several vessels may have to be used at once, depending on the complexity of the individual dishes and how many steps of cooking are necessary to complete the meal. Using more than one big pot at a time is practically impossible with a reflector cooker. Considering the variety of dietary patterns, heating and eating and cooking habits and the number of people who may wish to eat together, it is no wonder that solar cookers are widely regarded as too small. A good solar cooker should have adequate loading capacity in addition to some extra "holding space" for pots, pans and dishes.

Type of vessel: Kenyan and Malian women use different sized pots made of clay or aluminum. Aluminum pots are also the rule in Pakistan. If solar cookers will accommodate the customary vessels - instead of only pots with particularly flat bottoms (as they are necessary for the Stoy/Pöhlmann cooker), they will find a higher degree of acceptance.

Cleaning: Some solar cookers seem to be difficult to keep clean, e.g. after a pot has boiled over or spilled. For example, the Sobako cooker used in Kenya had to be partially dismantled for cleaning. Conversely, solar cookers that can be kept clean by wiping with a moist cloth are naturally more acceptable. Cleaning always comprises the danger of damaging the reflecting/absorbing surfaces. Here, too, solar cookers have to compete with conventional devices that are either easy to keep clean or rarely require cleaning.

Versatility: Box-type solar cookers and solar steam cookers are unsuitable for roasting and grilling and only conditionally suitable for baking. In fact, genuine versatility in the sense of being suitable for cooking, roasting, baking, frying, etc. is displayed by very few solar cookers - meaning that they are only capable of assuming certain limited functions in connection with the preparation of food.

Useful time: With the notable exception of the usually quite expensive and complicated heat-accumulating devices, solar cookers are only useful during certain hours of the day and seasons of the year, in addition to needing good weather.

Substitution of traditional cooking facilities: Many of the above considerations document the fact that solar cookers can rarely do more than supplement the existing cooking options, i.e. they cannot replace the traditional cooking facilities. Such was the case in India's solar cooker project in Gujarat. Traditional and solar-powered stoves/cookers are almost invariably used simultaneously, e.g. with the solar cooker serving to heat the food up, and the wood fire doing the rest of the cooking. As long as the traditional cooking facilities are also regarded as a source of heat and light, solar cookers can only be a supplement, never a substitute.

3.9 Solar Cooker Technology

Some of the more technical aspects of solar cookers constitute barriers to social acceptance. Some related problems that have occurred in past solar cooker projects are dealt with below.

Thermal output: The thermal output of a solar cooker includes all of the following:

- Efficiency: This is the ratio between the available heat and the incident energy, with the available heat being determined by heating a certain amount of water from a certain initial temperature up to the boiling point, and the incident energy being the measured global radiation times the aperture of the cooker for the same period of time (cf. section 2.2). In the case of heat-accumulating solar cookers, only the partial efficiency of the energy-accumulating components (collector and heat store without the cooking vessel) can be directly measured and therefore treated as a known quantity. There is an extensive lack of valid, comparable data on solar-cooker efficiency from methodically indubitable investigations. Consequently, the criteria applied here are more of a "social" nature: a solar cooker should, within a reasonable length of time, cook the staple foods of the target group,
- Temperatures: Solar cooking boxes and solar steam cookers achieve only moderate temperatures and take accordingly long to cook meals. On the other hand, the advanced-type reflector cooker for Mali reportedly develops excessively high temperatures;
- Temperature control: Most devices have no means of heat adjustment/control, though some - like most box cookers - have an indirect temperature-control option by way of the box and reflector alignment relativ to the sun's position.

- Starting power: Reflector cookers and the Solar Hot Plate Cooker have good starting power, but most others count as fairly weak;
- Temperature profile: Uniform heating of the pot is not always possible, depending on focus position, thermal contact, etc. The temperatures measured in various parts of a Sudanese cooking box, for example, showed considerable variance.

When one or more of the above aspects is given too little consideration at the design stage, the effect on acceptance can be devastating.

Durability: Solar cookers must be robust, sturdy and able to indefinitely withstand improper, if not to say grossly negligent, handling by the user and members of the family. The service life of a solar cooker must be much longer than its payback period.

Robustness: When one or the other part of a solar cooker has to be repaired due to minor damage, the whole cooker remains out of service until the part has been fixed. Usually, it is a broken mirror or pane of glass that impairs the cooker's serviceability. The fact that children, domestic animals and improper handling are quick to cause damage must be accounted for at the design stage.

Handling: Solar cookers that concentrate direct radiation at a focal spot require practice, a good grasp of the working principle, and constant close attention to the job at hand - namely handling and cooking. Some of the work involved consists of unaccustomed - sometimes even uncomfortable - manual exercises. Correct focusing of the incident radiation, continuous tracking of the sun, and working in and around the focal spot are among the "necessary evils". A solar cooker should not be evaluated on the basis of careful handling by trained personnel, but rather of somewhat careless handling by someone with less technical interest in the outcome. To put it more justly: by very busy people who have a lot of other things to worry about at the same time. Awkward, unaccustomed handling procedures - like extracting a hot, heavy pot - are an impediment to mass diffusion. In other words, ease of handling and convenient operation can decide whether or not an existing solar cooker will still be in use next month or next year.

<u>Solar tracking:</u> The need for solar tracking is one of the more tedious drawbacks of reflector cookers. Adjusting a Sobako 1 every 10 minutes is practically a full-time job all by itself. Depending on the situation, this necessity can be acceptable or unacceptable. The Sudanese solar cooking box has to be readjusted every hour or so. The rectangular shape of the ATRC, Serve, or Dhauladhar cookers makes solar tracking practically unnecessary. The need for constant, more or less complicated solar tracking is a major handicap for most reflector-type solar cookers.

<u>Glare:</u> Most people are irritated by the reflected sunlight in a reflector-type solar cooker. The Falco S/C scope of supply includes a pair of sunglasses. When stirring food or manipulating the cooking pot in the cooker's focal spot, one is exposed to blinding glare and possible injury by burning. That, too, is a substantial drawback of some reflector units. Some others, like the VIAX cooker, offer the possibility of easily turning the mirror out of focus for stirring and handling.

<u>Danger of injury:</u> Some solar cookers are still so primitive and inherently faulty constructed and/or finished (e.g. the Sudanese RERI Cooker), that they pose a real hazard to children who may be playing in the near vicinity.

<u>Breakdowns:</u> Some solar cookers, e.g. the Orangi Cooking Box, still require constant maintenance & repair work. That, of course, is not only detrimental in the long run, but also puts an unacceptable strain on the household budget of the average target-group family.

<u>Manufacturing materials:</u> If industrial components are used (panes of glass, aluminum foil, mirrors, special paints, plywood, ...) they will be hard to repair or replace with locally available materials. In India's Dhauladhar project, though, the brick & clay stoves were installed once and for all in the near vicinity of the user's house. The use of familiar

materials like clay, which is also used for building houses, facilitates production of the cookers - at the cost of mobility. The more local materials used, the greater the probability of long-term acceptance. Traditional hearths and stoves are practically always made of exclusively local materials.

Design: Add to the above a simple design and a well-worded, uncomplicated set of building instructions, and solar cookers can be built at low cost by native craftsmen and in small workshops or factories; repairs can be taken care of quickly and inexpensively. The simple design of the cooker used in North Horr, Kenya, for example, has earned much applause. The same is true of the mechanics of the Falco S/C Cooker. Obviously, the easier it is for people to build their own solar cookers, the more likely they will be to accept them. Good craftsmanship is necessary, though. Traditional-type hearths and stoves are often made by unskilled workers wielding a knife and hammer. Solar cookers, even the most simple cooking boxes, can hardly compete with that.

Defects: Perfect functioning of solar cookers is a main prerequisite for successful introduction. Technical deficiencies and faulty designs detract from the devices' utility value and make the target group lose interest before the project ever gets out of the trial phase.

- Take, for example, the (extremely inappropriate) Sobako cooker in Kenya, where the technical problems consisted mainly of:
  . inadequately secured heat-insulating panes around the oven
  . the matching vessels' tendency to rust
  . an oven that was 1 cm too short
  . swelling and distortion of various components under the effects of heat and humidity
  . uneven heating of the oven, and so on.
- A survey of 1572 households using ATRC cookers in Gujarat, India, revealed that the heat-absorbent paint used to coat the inside of the casing and the cooking vessels peeled off in 96 % of all cases. Additionally, 44 % of the users complained about steam collecting between the inner and outer

covers. The safety spacer between the outer cover and the reflector came loose on 41 % of the cookers. About 23 % of the cookers had broken or cracked covers. An equal share of households criticized the faulty seal and resultant air permeability between the box and the cover. The mirror was broken or damaged in 13 % of the cookers. Indeed, only 1 % of the cookers were found to be in perfect working order.

- The cooker used in Mali proved to have a major defect in the form of untimely formation of bubbles between the polyethylene foil and the reflector, which reduced the cooker's heat output. While it was no problem to get rid of the bubbles by punching holes in them and pressing the foil down against the background, the users nonetheless considered this to be a substantial disadvantage.
- Of 13 test families in the Sudan, 11 filled out and returned the survey questionnaire. Seven of them criticized the cooker's slowness, 6 its lack of heat control, 5 the need to cook outdoors, 3 the lack of accessibility, 2 the awkward cooking procedure and 1 the danger of eye injury posed by the reflector. Just for good measure, there were also jammed covers, warped boards, projecting screws, and generally inferior workmanship. Consequently, major improvements are envisioned: modified reflector, improved cotton insulation, more appropriate sizing, simpler construction, properly closing cover, etc. Once the causes of complaint have been eliminated, another technical test phase will have to be gone through before market studies can be conducted.
- Conclusions: Poor people cannot be expected to accept so many serious defects. Solar cookers must be technically mature before they are used in a field test.

Miscellaneous: Numerous other criteria are also of determining importance for acceptance, depending on the unit involved and the amount of experience gained: swelling of parts, defective insulation, heat losses, insulation modalities, etc. Only painstaking evaluation of individual solar cookers will show just how important such factors can be with regard to social acceptance. User-household surveys are a suitable starting point.

3.10 Economic Efficiency

The economic efficiency of a solar cooker is reflected by the time it takes to recoup the initial outlay and day-to-day maintenance costs by reducing commercial fuel consumption. Of course, the equipment must also remain serviceable substantially beyond its payback time. On the other hand, a purely monetary approach to the question of economic efficiency presents problems when applied to low-income households in which most of the fuel used is of noncommercial origin, e.g. cow pats, wood, scrub brush, etc. or to subsistence farming families with little or no involvement in the money economy. In the world's poorer countries, such families are often in the majority. The following aspects are of special importance.

Cost of production: For most solar cookers, little data is available on the actual cost of production. Since most of those in question are prototypes that do not yet display the technical maturity needed for series production, pertinent information is of low indicative value. Due to the chronic shortage of foreign currency in the Third World, preference should be given to cookers that can be made locally using indigenous materials. Whenever production costs are mentioned in connection with a solar cooker project, e.g. the Sobako project or the Sudan's Special Energy Program, they tend to go beyond the means of the target group. In relatively successful projects (Gujarat, SERVE), the sales price of the cookers is subsidized.

Purchase price: Practically any amount of money, however small, would still be too expensive for most rural households as long as firewood can be gathered for free (like in Mali) and the farmers earn very little money. A US$ 100 price tag would be utterly unacceptable. A family with a lower middle-class income (craftsman or laborer) in a town in Mali would have to pay between two and three months' earnings for a solar cooker. Even if the payback period were only 1 1/2 years (in towns), such a purchase would still amount to a major investment that only few could aford. Indeed, some 80 % of all urban families are unable to buy enough firewood for several days at once. In Pakistan, a single cooker of limited serviceability costs 2 months' pay. In Kenya, too, the in principle low cost of US$ 25 was still too high for most families, considering the meager income of the average rural household, coupled with the fact that higher priority has to be attached to other expenditures. Experience gained in other projects aimed at disseminating firewood-saving stoves in Kenya shows that a new stove should not cost more than US$ 10. Due to the extremely low purchasing power of the "needy poor", most solar cookers are bought by the comparatively "well-to-do poor". In India, the solar cookers proved too expensive for most people, even though the price was subsidized by 50 %. While the payback period of 2 to 3 months (for those who buy fuel) is really very short, the price of a solar cooker amounts to an enormous outlay for a low-income family. Consequently, most solar cookers are owned by people with medium-to-high incomes. But some such families still would have been willing to buy the cooker at the normal, unsubsidized selling price. Subsidized or not, solar cookers almost always prove to be too expensive for the target group. Additionally, due to their particular time preference, the needy always compare solar cooker prices with those of other kinds of stoves and cookers. As long as a traditional stove in the Sudan costs about 2 % as much as a solar cooker, the latter's competitive power will remain practically negligible among very poor social groups. To make matters worse, the cost of a solar cooker in the Sudan included a profit margin of as much as 600 %! While the envisioned market price for a Sudanese solar cooker stands at US$ 80, the average Sudanese peasant is willing to pay US$ 2...4 for an improved cookstove.

Cost of maintenance: In general, solar cookers have a reputation for needing little maintenance. By contrast, though, the aforementioned complaints about defective solar cookers indicate that the cost of maintenance can sometimes be substantial. Realistic repair-cost estimates have only been drawn up for a few models.

Savings: Assuming low maintenance costs, the economic efficiency of a solar cooker can be roughly estimated by subtracting the cost of procurement and maintenance from the cumulative savings on fuel. Since solar cookers can only be used at certain times of the day in certain seasons of the year, they can never replace a traditional hearth. Reports from Mali peg fuel savings at up to 50 %, and the users of India's ATRC cooker claim it reduces energy consumption by 1/3 to 2/3 and firewood consumption by 30...60 %. According to information from the Sudan, solar cookers used together with traditional hearths cut charcoal consumption by about 25 %. All things considered, just how much money actually can be saved by reducing fuel consumption depends on a number of factors:

- the kind of fuel being replaced
- the local and seasonal fuel price structure
- the number and nature of meals/dishes being prepared
- how many days a year a solar cooker can be used.

In general, solar cookers can be expected to reduce fuel consumption by 30...60 % in tropical and subtropical areas. According to one Indian study, 32 % of all families that use their solar cooker every day claim to be saving no fuel at all, probably because the simultaneous use of other cooking fires continues undiminished. Some 68 % of the daily users confirm that they are spending less on commercial fuel: 50 % save between US$ 1 to US$ 1.50 per month; more than 7 % say they save between US¢ 80 and US$ 1, and more than 42 % report savings of less than US¢ 70. Compared to the stated monthly income of the respondents (US$ 90...300) such savings are regarded as very modest. Table 10 lists some details. No opportunity costs have been calculated for time saved, but rural - underemployed or unemployed - laborers would probably attach a rather low value to it.

To the extent that the widespread use of solar cookers could appreciably retard the extraction of natural biomass for use as fuel, the ecological benefits would be quite substantial, though it would be hard to attach a monetary value to the conservation effect. On a global scale, however, progressive deforestation is due less to people's gathering firewood than to homesteading, the expansion of agricultural acreage, the lumber industry and industrial consumption.

Table 10: Achievable savings stated for the use of a solar cooker, Gujarat, India

Monthly savings	Urban and fringe areas	Rural areas	Total
less than US$ 0.7	37.00%.	63.41%	42.64%
US$ 0.8...1	53.80%	28.05%	7.01%
US$ 1...1.5	9.20%	8.54%	50.45%

Payback time: Table 11 shows how long a solar cooker would probably take to pay for itself at a purchase price of US$ 5...150, assuming roughly 45 % savings on fuel, with the original cost of fuel running at US$ 2.5...5 per family. Any cooker that will pay for itself in less than a year stands a good chance of acceptance in the sense of economic viability. Assuming that the 45 % reduction in fuel consumption is spread evenly over the entire year, and that there will be about 3 months worth of cloudy days per year, then the cutback in fuel consumption amounts to about 60 % during the 9 months of sunshine. That shortens the payback periods of less than 1 year (calculated on a year's-average basis) to about 3/4 of their listed duration (assuming that the entire payback period begins and ends in a single sunny season, because the average monthly savings during the dry season come to 60 % instead of the 45 % calculated for the full year). In such cases, then, the payback periods listed in table 11 can be multiplied by a factor of 0.75. Figure 12 lists the corrected payback periods for the cases in which amortization in a single dry season is possible. Obviously, this approach cannot be applied to tropical regions with two rainy seasons each year - with the possible exception of extremely favorable cases involving payback periods of 3 months or less.

Table 11: Payback period (in months) for a solar cooker, as a function of the purchase price and original monthly cost of fuel

Monthly expenditure for fuel	Purchase price of solar cooker						
	$ 5	$ 10	$ 25	$ 50	$ 75	$ 100	$ 150
$ 2.50	4.4	8.9	22	44	67	89	133
$ 5.00	2.2	4.4	11	22	33	45	67
$ 10.00	1.1	2.2	5.6	11	17	22	33
$ 15.00	0.7	1.5	3.7	7.4	11	15	22
$ 20.00	0.6	1.1	2.8	5.6	8.3	11	17
$ 25.00	0.4	0.9	2.2	4.4	6.7	8.9	13

Table 12: Payback period (in months) for a solar cooker during one single dry season, as a function of the purchase price and original monthly cost of fuel

Monthly expenditure for fuel	Purchase price of solar cooker						
	$ 5	$ 10	$ 25	$ 50	$ 75	$ 100	$ 150
$ 2.50	3.3	6.7					
$ 5.00	1.7	3.3	8.3				
$ 10.00	0.8	1.7	4.3	8.3			
$ 15.00	0.5	1.1	2.8	5.6	8.3		
$ 20.00	0.4	0.8	2.1	4.2	6.2	8.3	
$ 25.00	0.3	0.7	1.7	3.3	5.0	6.7	10.0

Payback periods amounting to more than 1 year for solar cookers as "objects of investment" are not regarded as negotiable, much less motivating, for the average household. This need not necessarily apply to large units intended for use in communal kitchens, in which case somewhat longer payback periods may appear acceptable. In general, though, only relatively short payback periods have a chance of being accepted, because the vast majority of people belonging to the social groups for which the procurement of fuel is a major cost factor, have very limited capital resources.

Additionally, in contrast to other kinds of renewable energy equipment with longer payback periods, such quick redemption of the capital outlay helps keep the interest burden (if any) at a minimum. In practice, no case is known of in which solar cookers were disseminated via interest-bearing loans. At the same time, families with low-to-medium incomes in developing countries hardly need fear any substantial loss of interest as a result of investing in a solar cooker. The tabular comparison of solar cookers (cf. appendix 1) does not include the payback period, because it is heavily dependent on local and seasonal fuel prices, and they, in turn are usually unknown. In addition, the payback period has no effect on how useful a particular type of solar cooker would or would not be in some other region.

Lifetime energy price: The lifetime energy price, PEL, is the purchase price of the solar cooker, P, plus expenditures for repairs/maintenance, R, multiplied by the useful life, L, of the unit, then devided by the useful energy produced by the cooker. The latter derives from the useful life, 270 days of service per year (9 months), the aperture (A, collector area) of the cooker, the cooking time, t, the mean global irradiance, G (at the time of cooking), the cooker's efficiency, η, and the number of meals cooked per day, n:

$$PEL = \frac{P + R \cdot L}{L \cdot 270 \cdot A \cdot t \cdot g \cdot \eta \cdot n} \quad \text{in US\$/kWh} \qquad (11)$$

The term "efficiency" is something of a problem for solar cookers, which also complicates calculation of the lifetime price of energy. For lack of any more accurate data, the efficiency values used here have been taken from the appended tabular comparison (calculated from the heatup power stated in the questionnaires (cf. section 2.2)). The average global irradiance, g, at the time of cooking was assumed to be 750 W/m^2. The same value was chosen for reflector cookers. The drawback of reflector cookers, namely that they only exploit direct radiation, is partially compensated for by their specific advantage, namely that they can be made to accurately track the sun during the cooking process. The cooking time is defined as 1.5 h for reflector cookers and 3 h for solar cooking boxes. As to the number of meals per day, n, it was postulated that between 1 and 2 meals are cooked in a solar cooking box, and a full two meals per day in a reflector cooker, which has substantially shorter cooking times. Thus, n = 1.5 for a box-type solar cooker, and n = 2 for a reflector cooker.

In the case of heat-accumulating solar cookers, the device's overall efficiency cannot be determined on the basis of how long it takes to boil a certain amount of water. They gather energy all day long and release it to the food at high power within a relatively short length of time when needed. This different situation is accounted for by an alternative definition of the lifetime cost of energy, PEL, for heat-accumulating solar cookers:

$$PEL = \frac{P + R \cdot L}{L \cdot 270 \cdot G \cdot 0.67 \cdot \eta' \cdot A} \tag{12}$$

This time, η' is the efficiency of the energy-collecting components (collector and storage), with no regard for heat transmission to or losses by the food. Presumably, the average total efficiency amounts to roughly 2/3 of the efficiency η' of the energy-collecting components, measured during a period of good insulation - thus the factor 0.67. We are fully aware of the fact that not only the calculation of efficiencies, but financial calculations in general, involve a substantial degree of uncertainty - no matter how intimidatingly precise the formulae may look. For example, few empirical data are available with regard to useful lifetimes and the cost of repairs/maintenance. Additionally, the efficiency data were not determined by neutral parties, and the test conditions were not uniform. Lastly, solar radiation data and their time histories differ widely from region to region.

<u>Cost of cooking:</u> The cost of cooking (CC), which is the specific cost of energy for cooking food during the first year, is relatively unencumbered with such uncertainties. While one could also relate the specific cost of energy to the entire service life of the cooker, the first year is the financially decisive year with regard to motivation in connection with the dissemination of inexpensive solar cookers in developing

countries. The cost of cooking during the first year is calculated according to the formula:

$$CC = \frac{P}{270 \cdot m \cdot n} \tag{13}$$

with m as the mass (weight in kg) of the food that can be cooked at once. Based on the first year of service, the economic-efficiency criterion CC clearly favors inexpensive appliances with a payback period of less than one year. Box-type solar cookers are the best bet in that sense. Reflector cookers with a PEL that shows they are no more expensive to operate than a solar cooking box are not as good at exploiting the incident energy. They do catch more energy than a cooking box, but they can't cook more food with it.

The lower energy yield, coupled with a higher purchase price, makes the CC of a reflector cooker significantly higher than that of a solar cooking box. Consequently, reflector cookers are financially unattractive for most families in developing countries.

Heat-accumulating solar cookers with a high purchase price have an economically unjustifiable CC. Their purchase price entails at least medium-term financial planning - and that can hardly be expected of most families in developing countries.

Consequently, in the case of heat-accumulating solar cookers, the cost of cooking CC has been replaced by the cost of cooking with respect to lifetime CCL, i.e. the price of energy per kg of food cooked over the entire useful lifetime of the cooker:

$$CCL = \frac{P + L \cdot R}{270 \cdot m \cdot n} \tag{14}$$

This has important consequences for the dissemination of heat-accumulating solar cookers: In practice, long-term financing and amortization mean that such devices are much more suitable for use by institutions like schools, hospitals, etc. than by private families, and then only if their capacity is large enough to justify the high purchase price.

<u>Summary:</u> In considering the economic efficiency of solar cookers, one must proceed on the basis of the very low real, average income of the majority of the population and take into account their fundamentally different time preferences. Otherwise, any conclusions drawn will tend to be false.

3.11 Direct and Indirect Impacts

Essentially indirect impacts, effects, side effects, aftereffects and consequential effects are sometimes accentuated in an attempt to justify solar cooker projects.

<u>Saving of time:</u> Solar cooking dispenses with a major share of the tedious, time-consuming work of gathering firewood, thus taking some of the load off the shoulders of the responsible women, men or children. In a box-type solar cooker, the food will not burn, needs no stirring, and therefore does not have to be watched. That, too, saves considerable amounts of time and energy. In Chinese reflector cookers, some dishes need shorter cooking time than on a conventional fire. In rural communities, time saved can be worth more than money saved. Ultimately, the social appreciation of saved time will depend on the woman's role and whether or not the time is saved at the cost of conversation. The macroeconomic importance of any time saved depends on its opportunity costs. Time savings should be evaluated on a socioeconomic - not economic - basis.

<u>Less smoke:</u> Solar cooking keeps smoke out of huts and is therefore seen as an especially clean way of cooking. Here, too, one must examine such arguments from a socioeconomic standpoint. While smoke from a cooking fire does have unhealthy effects, it also helps preserve food (which is intentionally stored in the same room). At the same time, it strengthens thatched roofs and makes them more water-repellent. Additionally, smoke keeps insects away and the fire serves as a source of light and heat at night. A solar cooker offers no such advantages.

Health effects: Solar cookers have multiple health effects:

- air quality: The absence of smoke at the place of cooking means better air quality for women who used to cook over a wood or charcoal fire. In densely populated areas, that advantage is enjoyed by the population in general;
- burns: Reflector cookers pose the danger of burning/blinding. On the other hand, conventional hearths and stoves are much more dangerous for children;
- circulatory stress: Reflector cookers force the cook to stand out in the hot sun. In Mali, especially the pregnant and nursing mothers complained of frequent headaches and dizziness. Here, too, the pros and cons have to be carefully balanced on a case-by-case basis, i.e. generalizing is uncalled-for.

Nutritional effects: Solar cooking boxes cook food gently, thus preserving many nutrients and flavor substances that would normally be lost by cooking over a fire. On the other hand, it has not yet been - but should be - clarified to which extent slow cooking - particularly in box-type solar cookers, where the food is held at relatively low temperatures for considerable long time - could lead to decomposition with possibly toxic consequences, specially in aluminum vessels. Applied nutritional-physiological research would be appropriate in that connection.

Effects on balance of payments: Solar cooking saves fossil fuels like petroleum and bottled gas. For countries with a lack of foreign exchange and no indigenous oil or natural gas, that translates into less dependence on imported energy and resultant disburdening of the balance of payments. However, since solar cookers are still so sparsely disseminated, with only few poor target groups actually using them, it would hardly be worthwhile to even attempt to calculate such indirect consequential effects.

Ecological impact: The same applies to the consequential effects of reducing firewood consumption, namely a more sparing use of natural resources, preservation of the eco-system, less deforestation, desertification and soil erosion, groundwater protection, preservation of climate, avoidance of drought. The relative contribution of solar cookers to any of these is too minor to warrant any serious attempt at quantification. Contributions by other projects, programs and policies are of much greater significance, especially projects dealing with reforestation, massive dissemination of efficient wood-burning stoves in limited areas, control of lumber harvesting, and promoting efficient energy consumption in transportation and industry.

3.12 General Conditions of Acceptance

The decision to buy and use a solar cooker rests in part on considerations of a more general nature. A good many of the arguments discussed above can have a positive or negative bearing on the final decision.

Decision to procure: Asked why they bought a solar cooker, 1572 users in Gujarat, India, named the following motives /139/:
- savings on fuel (61 % of those interviewed)
- wish to try something new (21 %)
- time savings (8 %)
- subsidized purchase price (8 %)
- no more burned food (2 %)

In well over half of the households, the decision to buy a solar cooker was made by the male head of the family. Only one in six solar cookers was purchased on the basis of a common decision by husband and wife. In the Sudan, though, it is customary for the women to purchase household items with their husband's money; they have a say in decisions concerning innovation.

Utilization motives: In each different social group and for each different type of solar cooker, the entire network structure of acceptance conditions presents itself anew - and differently. The aforementioned survey of 1572 households in Gujarat, India, drew attention to the following problems concerning acceptance of the Indian Solar Cooking Box /139/:
- cooker too heavy (criticized by 37 % of those questioned)

- cooking takes too long (30 %)
- cooking time runs counter to the family's workaday routine (30 %)
- no suitable place to put the solar cooker (9 %)
- rejection of aluminum cookware (7 %)
- altered taste of food (4 %)
- complicated handling (4 %)

As a result of such problems, many families have stopped using their cookers. Indeed, of the 2204 families surveyed, approx. 1/3 (818) said that they still used their solar cooker, while 1006 said they still have their solar cooker but no longer use it. Some 226 families had either sold their solar cooker or given it away. A different survey of 490 households came to the conclusion that about 12 % of the cookers were in disuse /116/. More than 22 % of the families use them to cook one or two meals a week, and 66 % use them daily. The situation in Afghan refugee camps in Pakistan is similar: only 2 families out of 10 actually used their solar cooker for preparing meals; the other 8 families said they use their cooker rarely or not at all - for the following reasons:

- broken glass cover
- damaged mirror
- cooker too small for a large family
- cooking takes too long
- cooking on a solar cooker only possible during the summer months
- aversion to preparing/cooking meals in the sun
- unaccustomed to operating the cooker

One woman claimed to have no food to cook.

Awareness: The main prerequisite for buying and using a solar cooker is a subjective need, as opposed to objective necessity, for energy-saving devices. The real motivation is determined by the urgency and existential importance of the need in question as compared to other basic necessities. But motivation can also stem in part from a desire to decorate and furnish one's living quarters. Whether or not problem-consciousness exists, or is even possible, can only be ascertained by way of an exact situation analysis of the target group. The Sudanese Energy Development Plan, for example, calls attention to the fact that the people themselves are not yet aware that firewood is a problem, though the situation in the northern part of the country is much more serious than in the south. In India, where practically every family is affected by the increasing scarcity of fuel, awareness of the problem is still estimated as being very low. For example, cooking fires are often not promptly extinguished; in some families, they are left to burn for hours on end. Of course, a large share of rural families still gather their wood for free and therefore cannot be expected to attach existential importance to having a solar cooker. Many women say they would like to have a solar cooking device, but their other problems are more urgent: getting the day's food, worrying about the family's health, etc. As long as the target group is not personally affected, they will see little reason to buy and use solar cookers. Besides, one can always adopt an "alternative strategy" like eating fewer hot meals in favor of more bread, etc.

User structure: In hardly a solar cooker project are the actual users of the solar cookers identical with the needy target group envisioned by the project-executing organization. All solar cookers produced to date in connection with the ATDO project in Pakistan are in the possession of Karachi residents. These people bought cookers because they wanted to try something new, not because they were suffering from a shortage of fuel. Since the families in question belong to the middle and upper-middle classes, they were able to buy their devices at unsubsidized prices. Though India's official program aspires to achieve widespread use of solar cookers among low-income groups, the dissemination measures are aimed primarily at the more well-to-do social strata. Since the strategy is to give solar cookers a status-symbol image in the hope that poor families will try to emulate the rich by buying and using one, the result could be to awake artificial needs instead of engeneering an appreciation of the need to conserve energy. Indeed, how much of a percolating effect may or may not occur still remains to be seen. One thing is for sure, though, namely that the users rarely belong to the target group - who continue to use wood to meet their energy demands.

Willingness to accept change: At times, reference is made to the need for a willingness to accept change (as opposed to the ability to accept change). If a solar cooker is economic and fits neatly into people's dietary, eating and cooking habits, it will find acceptance. Poor people nearly always show more purposeful, shrewder behavior than other people do. Consequently, one should avoid the use of "traditional" and "conservative" (read: prejudiced) thought patterns, so as not to seem incapable of understanding the target group's everyday rationalities. On the contrary, they should be carefully investigated as the basis of a technology impact assessment for innovations like solar cookers.

3.13 Summary

The above observations, reflections and comments show that, while some good, serviceable solar cookers have been developed, no appreciable degree of acceptance by target groups has been achieved. The reasons behind the mediocre success of solar cooking are less of a technical nature than of a sociocultural, socioeconomic and psychosocial nature. Despite continuous technical improvement, it has not been possible to adequately adapt the devices to the real wants and needs of the users.

While solar cooking boxes are widely regarded as the most practical approach to solar cooking, experience shows that no attempt to achieve mass dissemination has yet been successful. Field studies conducted in Gujarat, India, revealed that even families who own a solar cooker do not always use it. While solar cookers do have their advantages, they do not fully accommodate the wants and needs of the general population. Reflector cookers have experienced substantial acceptance - if any - only in China so far.

The possibility of cooking with solar energy is nobody's secret. But most people still fail to realize that, in some areas, solar cooking may soon constitute one of the few remaining options for preparing a hot meal. On the other hand, test projects in India and China have shown that the dissemination of solar cookers actually is possible. The extent to which future projects will contribute to the further dissemination of solar cookers, will extensively depend on how much attention is paid to the socioeconomic, sociocultural and psychosocial criteria dealt with on these pages.

4. Questions Concerning Solar Cookers and Solar Cooker Projects

As indicated by the above considerations, there are many questions to be asked in connection with solar cooker projects. Those questions are listed below in the form of a questionnaire. The more affirmative answers given, the better the conditions for solar cooking. Only a detailed on-the-spot analysis will show just which of the following questions are the most essential:

Solar energy supply

1. Does the solar energy supply suffice to allow the near-constant utilization of solar cookers?
2. Is the solar energy supply adequately uniform and usually predictable?
3. Is the solar cooker able to utilize both direct and indirect/diffuse radiation?
4. Does the topographical situation of the target area favor the use of solar cookers?
5. Can the solar cooker be used from shortly after sunrise until shortly before sunset?
6. Are the reduced seasonal insolation levels (rainy season, wintertime, ...) still acceptable?

Energy demand

7. Is the energy consumption of the target population distinctly on the rise?
8. Are the needy in a precarious situation with regard to the availability of energy?
9. Does the target group consume disproportionate amounts of energy in their households (compared to other uses, e.g. for transportation)?

10. Is the procurement of energy for cooking an especially urgent social problem among the needy?
11. Does the target group need a conspicuously large amount of time to procure fuel for cooking?
12. Does the target group - compared to other social groups - consume disproportionate amounts of energy for cooking due to their particular situation, e.g. place of residence, local customs, ...?
13. Does relatively little household energy go for baking (bread, etc.)?
14. Is the cooking-energy price structure relatively high and/or on the rise?
15. Is most cooking done over a wood fire?
16. Is lots of charcoal used for cooking?
17. Do national energy policies promote the dissemination of solar cookers?

<u>Social situation</u>

18. Does the target group have enough buying power to procure and maintain a solar cooker?
19. Are solar cookers presently in use among sections of the population with a social multiplying effect?
20. Do the people's systems of values allow the use of solar cookers?
21. Would the use of a solar cooker improve the average housewife's situation, e.g. in the sense of time saved?

<u>Dietary patterns</u>

22. Does the target group prefer to boil (as opposed to frying, etc.) their staple dishes?
    a) 1st staple food: ______________________
    b) 2nd staple food: ______________________
    c) 3rd staple food: ______________________
23. Are the main meals and/or "national dishes" easy to prepare with solar cookers?
    a) 1st main meal: ______________________
    b) 2nd main meal: ______________________
    c) 3rd main meal: ______________________

24. Are the basic breakfast dishes easy to prepare on a solar cooker?
    a) 1st breakfast dish: ____________________
    b) 2nd breakfast dish: ____________________
    c) 3rd breakfast dish: ____________________
25. Do people eat their main meal, at midday?
26. Can the traditional methods of preparing the main meals be maintained when a solar cooker is used?
27. Does bread play a subordinate role among the main dishes?
28. Can the baked goods preferred by the target group be prepared in a solar cooker?
29. Can solar cookers be used to prepare more than just the side dishes?
30. Can the possibility of a solar cooker changing a food's flavor be ruled out?
31. Is the majority of the target group's food chopped up for cooking (as opposed to large chunks)?
32. Are the customary beverages easy to prepare on a solar cooker?
33. Is the time it takes to prepare, say, tea on a solar cooker considered acceptable?

Eating habits

34. Is breakfast eaten relatively late in the morning, i.e. long enough after sunrise?
35. Is the evening meal taken before sunset?
36. Is it easy to prepare a meal with several courses on a solar cooker?
37. Can menus for groups of substantial size, e.g. festivities, be prepared on solar cookers?

Cooking habits

38. Are the cooker and pots easily accessible?
39. Does using a solar cooker shorten the cooking time or at least not lengthen it?
40. Can cooking be interrupted without difficulty?
41. Does the solar cooking process require relatively little attendance in the way of stirring, etc.?
42. Can the cooking process and the remaining cooking time be checked without difficulty?

43. Can meals be prepared in a minimum number of steps?
44. Is the cooker suitable for keeping food warm over a considerable length of time?
45. Can burned food be ruled out?

Comparison with hearths, fireplaces and cookstoves

46. Is the solar cooker as versatile as the traditional cooking facilities?
47. Can meals be cooked at any time, night or day?
48. Does the solar cooker have positive side effects equal to those of traditional cooking facilities, e.g. heating, lighting, conversation?
49. Can the solar cooker be used at the same time of day as traditional cooking facilities?
50. Can the solar cooker also be used for baking, frying, etc. instead of only for cooking in the sense of boiling?
51. Does the solar cooker have an adequate food capacity (quantity)?
52. Does the solar cooker have an adequate loading capacity (space)?
53. Can traditional cooking vessels/commercial-type pots and pans be used on the solar cooker?
54. Does the solar cooker have a modest space requirement at its place of use?
55. Is the target group extensively willing to cook outdoors?
56. Can the solar cooker be set up quickly and easily?
57. Is the solar cooker adequately stable?
58. Can the housewife stand in the shade while cooking a meal?
59. Is the height of the solar cooker about the same as that of the locally traditional cooking facilities, so that cooking can be attended to in the customary position, e.g. sitting down?
60. Is the solar cooker easy to clean, e.g. without having to be dismantled and without damage to absorbing or reflecting surfaces?
61. Does using the solar cooker make it unnecessary to use traditional cooking facilities at the same time?

Solar cooker technology

62. Is the solar cooker easy to move and to handle?
63. Is solar tracking unnecessary?
64. Can the danger of blinding or dazzling be ruled out?
65. Can other injuries as a result of using the cooker be ruled out?
66. Can the danger of injury by burning be ruled out?

Quality of design & workmanship

67. Is the solar cooker of high quality, robust and durable?
68. Does it rarely need repair?
69. Can it be made and kept in good repair using locally available materials?
70. Is the solar cooker suitable for do-it-yourself construction and repair?
71. Have only few complaints been lodged in the past?
72. Is the solar cooker technically mature?
73. Is the solar cooker children-, animal-, and fool-proof and -resistant?

Performance

74. Is the thermal output adequate?
75. Is the efficiency high enough?
76. Are the temperatures appropriate?
77. Does the cooker have a high heatup power?
78. Is it equipped for temperature control?
79. Is a constant temperature achievable?

Economic efficiency

80. Is the solar cooker inexpensive to produce?
81. Is the solar cooker inexpensive to procure?
82. Is the purchase price acceptable to the target group?
83. Is the cost of maintenance acceptably low?
84. Can considerable amounts of fuel be saved?

Direct and indirect impacts

85. Does using the solar cooker yield substantial time savings?
86. Is reduced smoke development regarded as an advantage?
87. Does using a solar cooker improve the quality of air for breathing?

88. Can unhealthy effects of cooking in the sun be ruled out?
89. Are nutritional and flavor substances preserved by solar cooking?
90. Can toxic reactions due to lengthy cooking be ruled out?
91. Would the use of solar cookers yield more than just a token improvement in the balance of payments?
92. Can the use of solar cookers be expected to have significant/relevant ecological effects?

General conditions of acceptance

93. Do (men and) women decide (together) whether or not to buy a solar cooker?
94. Have other, comparable target groups not only bought solar cookers, but also used them for a considerable length of time?
95. Is the target group aware of the firewood situation?
96. Can (especially) needy target groups in particular be expected to use solar cookers?
97. Can the use of solar cookers by middle-class families or intermediary subgroups be expected to have substantial spinoff effects on the poorer sections of the population?
98. Does the target group demonstrate a willingness and capability to accept change?

5. References

1 Achaya, K.T. (ed.): Interfaces between Agriculture, Nutrition, and Food Science. Tokyo (UNU) 1984

2 Agricultural University in Collaboration with Sonnenkorb: Feasibility of Solar Cookers in Urban and Rural Areas. Research Project by Department of Home Management. Hyderabad 1984

3 Ahmed, A.S.: Religion and Politics in Muslim Society - Order and Conflict in Pakistan. New York 1983

4 Aprovecho Institute: Fuel-Saving Cookstoves. Braunschweig (Vieweg) 1984

5 Arbeitsgemeinschaft für Entwicklungsplanung: Solare Koch- und Backgeräte. I. Phase: System- bzw. Leistungsvergleich

von solaren Koch- und Backgeräten unter technischen und ökonomischen Gesichtspunkten. München 1978

6 Arbeitsgemeinschaft für Entwicklungsplanung: Solare Koch- und Backgeräte. II. Phase: Prototypenbau und Erprobung. München 1981

7 Arbeitsgemeinschaft für Entwicklungsplanung: Sonderenergieprogramm Mali. München 1981

8 Bahador Mehdi, N.: Conceptual Development of a Solar Town in Iran. In: Solar Energy, 23, 1, 1979, 17-

9 Bahadori, M.N.: Solar Energy Utilization for Developing Countries. In: Ekistics, 45, 269, 1978, 172-

10 Berg, J. van den: Village Technologies: Lightening the Work Burden of Women. In: Afrika, German Federal Republic, 21, 2/3, 1980, 28-30

11 Bernard, R.: A Handy Solar Cooker. In: Sunworld 11, 2, 1987, 50-51

12 Bernard, R.: Easy to Build Solar Cookers. In: Solar Age, 3, 2, 1978, 14-

13 Bezbaruah, A.N.: Experiments with Low-cost Solar Devices such as Solar Cookers and Solar Water Heaters, and Bio-Gas Plants with an Aim for Better Utilization of Solar-Energy in the Northeastern Indian State of Assam. In: [175, 401-411]

14 Brattle, L.V., Irving, R.J.: Energy Needs for Cooking in the Sudan - An Interdisciplinary Approach to the Domestic Fuel Problem. In: International Solar Energy Congress, Brighton, England, Aug 23-28, 1981

15 Brattle, L.V., Irving, R.J.: The Use of Solar Energy for Cooking in Developing Countries. In: Journal of Consumer Studies and Home Economics, 10, 3, 1986, 261-270

16 Brown, N.L. (Hrsg.): Renewable Energy Resources and Rural Applications in the Developing World. Boulder (Westview) 1978

17 Bruchhaus, E.M. u.a.: Frauen in Mali - Situationsanalyse und entwicklungspolitische Ansatzpunkte unter besonderer Berücksichtigung nicht-staatlicher Organisationen. Freiburg 1979

18 Bruchhaus, E.M.: Herdbauaktivitäten in Kenia. Eschborn (GTZ) 1984

19 Brück, E., Porz, R., Scheffler, W., Swillus, O.: Sonnenherd: Leistungsstark, einfach, kostengünstig. In: Sonnenenergie 9, 5, 1984, 4-6

20 Bundesministerium für Ernährung, Landwirtschaft und Forsten: Schwerpunkte der Internationalen Agrarforschung. Münster-Hiltrop 1986

21 Bundesministerium für Forschung und Technologie: Neue Technik für die Dritte Welt. Bonn, without year

22 Bundesministerium für wirtschaftliche Zusammenarbeit: Programm der Bundesregierung für die Zusammenarbeit mit Entwicklungsländern auf dem Gebiet der Energie. Bonn 1983

23 Bundesministerium für wirtschaftliche Zusammenarbeit: Planung und Steuerung von Vorhaben der bilateralen Finanziellen und Technischen Zusammenarbeit. Bonn 1985

24 Bundesministerium für wirtschaftliche Zusammenarbeit: Grundlinien der Entwicklungspolitik der Bundesregierung. Bonn 1986

25 Bundesministerium für wirtschaftliche Zusammenarbeit: Aus Fehlern lernen. Bonn 1986

26 Bundesministerium für wirtschaftliche Zusammenarbeit: Hauptbericht zur Evaluierung des Sonderenergieprogramms Sudan (SEP). Entwurf. Bonn 1986

27 Bundesstelle für Außenhandelsinformationen: Indien. Energiewirtschaft 1983/84. Köln 1984

28 Bundesstelle für Außenhandelsinformationen: Kenia. Energiewirtschaft 1983/84. Köln 1984

29 Bundesstelle für Außenhandelsinformationen: Pakistan. Energiewirtschaft 1982/83. Köln 1984

30 Buxey, D.: Solar Cookers. In: [172, 1762-1767]

31 Cecelski, E.: Energy and Rural Women's Work: Crisis, Response and Policy Alternatives. Geneva (ILO) 1986

32 Cecelski, E.: The Rural Energy Crisis. Womens's Work and Basic Needs: Perspectives and Approaches to Action. Geneva (ILO) 1985

33 Christensen, H.: Afghan Refugees in Pakistan: From Emergency towards Self-Reliance. Geneva 1985

34 Christensen, H.: Sustaining Afghan Refugees in Pakistan. Geneva 1983

35 Consortium on Rural Technology: Solar Cookers and Ovens. Technology Options. New Delhi 1982

36 Cope A.W.G., Tully, N.: Considerations on the Development of an Urban Solar Cooker. In: South African Mechanical Engineer, 31, 9, 1981, 226-234

37 Dang, A.: An Analytical Study of a Solar Cooker Augmented with a Booster Mirror, Using PCM as Storage. In: Energy Conversion and Management, 25, 3, 1985, 255-261

38 Datta, R.L.: Solar Energy Utilization in Developing Countries. In: Ekistics, 38, 225, 1974, 93-

39 Dayall, M.: Food-Energy Nexus Activities in India. Tokyo (UNU) 1985

40 Department of Non-Conventional Energy Sources (DNES), Ministry of Energy: The Layman's Guide to Solar Cookers. Cooking the Natural Way. New Delhi 1984

41 Deutsche Gesellschaft für Technische Zusammenarbeit (GTZ): Herdverbreitung - ein Konzept der angepaßten Technologie. Sektorheft "Aus der Arbeit der GTZ", Nr.3, 1986

42 Deutsche Gesellschaft für Technische Zusammenarbeit (GTZ): Solar Cooker Kenya - Final Report. Eschborn (GTZ)

43 Deutsches Zentrum für Entwicklungstechnologien (GATE): Frauenförderung in der Technischen Zusammenarbeit. Situation, Ziele, Konzeption und Instrumente. Without place and year.

44 Deutsches Zentrum für Entwicklungstechnologien (GATE) (Hrsg.): Survey on Solar Cookers. Eschborn (GATE) 1978

45 Donner-Reichle, C.: Die Last der Unterentwicklung - Frauen in Kenia. Berlin 1977

46 Dunkui, M.: Solar Energy Applications in Remote Areas of Northwestern China. In: Sunworld 10, 2, 1986, 46-51

47 Eberhard, A.A.: Dissemination of Solar Ovens in Lesotho - Problems and Lessons. In: [164, 2754-2758]

48 Egneus, H. et al. (Hrsg.): Bioenergy 84. Barking (Elsevier) 1984

49 Elshorst, H.: Entwicklungspolitik. Bilanz und Perspektiven. In: Entwicklungspolitik epd 9/10, 1986, 31-50

50 Energy Research Council & Sudan Renewable Energy Project: Improved Charcoal-Stoves for the Sudan. A Guide for Introducing Technologies. Khartoum 1986

51 Erlbeck, R., Sepp, C.: Herdleitfaden - Leitfaden für Programme zur Verbreitung brennholzsparender Herde. Eschborn (GTZ/GATE) 1985

52 Fang, P.H., Swannack-Nu, S.: Chinese Solar Cookers. In: Appropriate Technology, 5, 4, 1979, 4-5

53 FGK-Unternehmensberatung: Beratung der Frauenorganisation Maendeleo ya Wanawake bei der Verbreitung von Geräten zur Nutzung nicht-konventioneller Energieträger. Eschborn (GTZ) 1980

54 French, D.: The Economics of Renewable Energy Systems for Developing Countries. Norfolk (AID) 1979

55 Garg, H.P., Bandyopadhyay, B., Datta, G.: Mathematical Modeling of the Performance of a Solar Cooker. In: Applied Energy, 14, 3, 1983, 233-239

56 Garg, H.P., Mann, H.S., Thanvi, K.P.: Performance Evaluation of Five Solar Cookers. In: [179, 1491-1497]

57 Garg, H.P., Thanvi, K.P.: Simple Hot Box Type Solar Cooker. In: Research and Industry, 21, 3, 1976, 184-186

58 German Appropriate Technology Exchange (GATE) (ed.): Survey on Solar Cookers. Eschborn 1978

59 Ghrab-Morcos, N.: Cours d'énergie solaire. Tunis (ENIT) 1984

60 Government of India: Seventh Five Year Plan 1985 - 90. Vol.1,2. New Delhi 1985

61 Government of Pakistan: Oxford Atlas für Pakistan. Karachi 1985

62 Government of Pakistan, Planning Commission: Special Development Plan for Tribal Areas of the North West Frontier Province. Islamabad 1982

63 Gujarat Energy Development Agency (GEDA): Evaluation of Prototype Solar Cooker Designs. Vadodara 1986

64 Gupta, R.: Solar Ovens (Lack of Acceptance in India). In: World Agriculture, 32, 3/4, 1983, 4-

65 Guzman, M.T., Jury, G.: Contribution de la mujer, especialmente en zonas rurales a la adaptacion, uso y desarrollo de nuevas tecnologias: Concurso cocinas solares. Informe final. Tomo I: Antecedentes, desarrollo y proyecciones. Tomo II: Documentos y anexos. Santiago (UNESCO, ROSTLAC, INTA) 1987

66 Halacy, B.: The Solar Cookery Book: Everything Under the Sun. Culver City (Peace) 1978

67 Hall, D.O., Morton, J. (eds.): Solar World Forum. Solar Technology in the Eighties. Oxford (Pergamon) 1982

68 Hall, C.A., Swet, C.J., Temanson, L.A.: Cooking with Stored Solar Heat. In: [179, 547-553]

69 Hamadto, S.A.: Planar - Sectioned Solar Concentrators. 1: Polygon Reflectors. In: Applied Optics 23, 8, 1984, 1279-1283

70 Hamadto, S.A., Al-Shibani, Khider, M.: Simple Geometry of Solar Concentrators. In: Sunworld 7, 3, 1983, 68-70

71 Hauptberatungsstelle für Elektrizitätsanwendung (Hrsg.): HEA-Bilderdienst, Heft 6.2: Elektroherd. Frankfurt 1985

72 Hayes, D., Kadyszewski, J.: Third World Energy Options and the Role of the Solar Energy Research Institute. In: Natural Resources Forum, 5, 2, 1981, 141-

73 Hoda, M.M.: Solar and Bulb Cooker. In: ISES/Solar Energy Society of Canada Energex 8th Conference Regina, Canada, Aug 23-29, 1982, 288

74 Hoda, M.M.: Solar Cooker. In: Appropriate Technology, 4, 2, 1977, 28-29

75 Hoda, M.M.: Solar Cookers. ATDA Research Report Series 1. Lucknow 1980

76 Hughart, D.: Prospects for Traditional and Non-Conventional Energy Sources in Developing Countries. Washington (IBRD) 1979

77 Hühn, H.G., Reuber, M., Forsbach, H., Suding, P.H.: Energy Survey on Family Level in Afghan Refugee Tentage Villages in Pakistan. Köln 1986

78 Indian Association for the Advancement of Science, Department of Non-conventional Energy Sources: Cost-Benefit Analysis: Solar Water Heating System and Solar Cooker. New Delhi 1986

79 Interdisciplinäre Projekt Consult (IPC): Special Energy Programme Kenya - Renewable Energies in Northern Kenya. Eschborn (GTZ) 1983

80 Intermediate Technology Development Group (ITDG) (ed.): Technical Brief: Solar Cooking. Rugby, Warwickshire 1983

81 International Bank for Reconstruction and Development (IBRD): Sudan: Issues and Options in the Energy Sector. Washington 1983

82 International Labour Office (ILO) (ed.): International Workshop on the Rural Energy Crisis, Women's Work and Basic Needs. Den Haag, 21.-24.4.1986. Geneva 1986

83 Ismail, E.: Women and Development. Energy Consumption: Household Survey in Two Gezira Villages. Khartoum 1987

84 Jain, A., Bhat, C.M.: Acceptability and Vitamin Contents of Foods as Affected by Cooking in a Sauce Pan, Pressure Cooker and Solar Cooker. In: Journal of Food Science and Technology-Mysore, 21, 4, 1984, 227-230

85 Janzen, A.F., Swartman, R.K. (eds.): Solar Energy Conversion II. Toronto (Pergamon) 1981

86 Jerman, B.: Lightweight Portable Solar Cooker. In: [67, 601-606]

87 John, V., Jagadeswaran, R., Satyanarayana, K.G., Mathew, A.G.: A Solar Cooker for the Common Man. In: Research and Industry, 30, 1, 1985, 1-5

88 John, V., Madhava, M.R., Rohatgi, P.K.: Solar Cooking of Cassava. In: Energy, 7, 3, 1982, 313-317

89 Kandpal, T.C., Mathur, S.S.: The Economics of Box-Type Solar Cookers. In: Energy Conversion and Management, 26, 2, 1986, 233-235

90 Kapstein, E.B.: The Solar Cooker. In: Technology and Culture, 22, 1, 1981, 112-121

91 Kenya Fuelwood Project: Energy Development in Kenya - Problems and Opportunities. Nairobi 1982

92 Khalifa, A.M.A., Taha, M.M.A., Akyurt, M.: Design, Simulation, and Testing of a New Concentrating Type Solar Cooker. In: Solar Energy, 38, 2, 1987, 79-88

93 Khalifa, A.M.A., Taha, M.M.A., Akyurt, M.: Solar Cookers for Outdoors and Indoors. In: Energy, 10, 7, 1985, 819-829

94 Kießling, G.: Erfahrungsbericht über den Einsatz des Sobako 1 (Sonnenbackofen) in Yaounde (Kamerun). Eschborn (GTZ) 1979

95 Kleinwächter, H., Kleinwächter, J.: Fix-Fokus Konzentrator zum Betrieb einer stationären Sonnenküche. In: Hohmann, H. et al. (eds.): Tagungsbericht des 4. Internationalen Sonnenforums. Berlin 6.-9.10.1982. München 1982, 479-485

96 Kühnert, W.: Sonnenback- und Kochgerät Kenia, Durchführungsplan. Eschborn (GTZ) o.J.

97 Krosigk, H.: Verbreitungsprogramm für holzsparende Herde in ländlichen Gebieten (Kenia) - Abschlußbericht der Technischen Beraterin. Eschborn (GTZ) 1986

98 Kuhnen, F.: Landwirtschaft und anfängliche Industrialisierung. Opladen 1968

99 Kupfernagel, R., Schaffernicht, R.: Latentwärmespeicher für solare Koch- und Backgeräte. Berlin (IPAT, GATE) o.J.

100 Lateinamerikanisches Institut: Erfahrungen mit dem Einsatz von Sonnenkochern in Entwicklungsländern. St. Gallen (SVMT/ILE) 1976

101 Lawand, T.A.: The Potential of Solar Cooking in Developing Areas. In: [170, 133-136]

102 Leach, G.: Residential Energy Use in Pakistan. Highlights on the 1979 Survey. o.O. 1984

103 Lillywhite, M., Lillywhite, L.: Can Passive Solar Technology Help Meet African Village Energy Needs? In: Passive Solar 5th National Conference, Amherst, Oct 19-26, 1980, 51-

104 Lof, G.O.G.: Solar Energy in the Less Developed Countries. In: [16, 27-43]

105 Malathi, D., Swaminathan, K.R., Kamaraj, S., Palanisamy, P.T.: Utilisation of Biogas for Household Cooking in India. In: [48, 36-39]

106 Malhotra, K.S., Nahar, N.M., Rao, B.V.R.: An Improved Solar Cooker. In: International Journal of Energy Research, 6, 3, 1982, 227-231

107 Mani, A. (ed.): Handbook of Solar Radiation. Data for India 1980. New Delhi 1981

108 Mannan, K.D., Cheema L.S.: Development and Testing of Solar Cooking Devices at P.A.U. Ludhiana. In: Alternative Energy Sources. 6th International Symposium, Miami Beach, Dec 12-14, 1983, 525-

109 Mekki, R.: Patterns of Household Energy Use in Sudan. Khartoum (Ministry of Energy and Mining) 1984

110 Mills, D.R., Qiu, M.Y.: New Solar Cooking Stove with Thermal Storage. In: Sunworld, 11, 2, 1987, 44-49

111 Ministry of Energy and Regional Development: National Energy Policy. Kenya. Draft 3. Nairobi 1985

112 Mock-Bieber, E.: Dienstreisebericht. Berlin 1984/85

113 Mock-Bieber, E.: Studie über ernährungs-physiologische Auswirkungen verlängerter Gar- und Wärmezeiten von Speisen. Berlin 1984

114 Monsted, M.: Women's Groups in Rural Kenya and their Role in Development. Copenhagen 1978

115 Morgan, W.B. et al. (eds.): Rural Energy Systems in the Humid Tropics. Tokyo (UNU) 1980

116 Moulik, R.: Socio-Psychological and Economic Factors Affecting Acceptability of Solar Cookers in Gujarat. Vadodara 1985

117 Nag, K.N., Mathur, A.N.: Performance Study of a Solar Cooker (Food Preparation, India). In: AMA, Agricultural Mechanization in Asia, Africa and Latin America, 14, 2, 1983, 69-71

118 Nahar, N.M.: Performance and Testing of an Improved Solar Water Heater Cum Steam Cooker. In: International Journal of Energy Research, 9, 1, 1985, 113-116

119 Nahar, N.M.: Performance Studies on Different Models of Solar Cookers in Arid Zone Conditions of India. In: [174, 431-439]

120 Nandwani, S.S.: Study of a Cheap Solar Oven in the Climate of San-Jose (Costa Rica). In: [85, 488]

121 National Council of Applied Economic Research: Report on Rural Energy Consumption in Northern India, Environment Research Committee, Department of Science and Technology. New Delhi 1981

122 National Energy Plan Committee: The National Energy Plan 1985 - 2000. Khartoum 1985

123 National Energy Administration: Sudan Energy Handbook. Khartoum 1987

124 Neubauer, L., Williams, G.: Solar Oven Economy for Farm Houses. In: Agricultural Engineering, 57, 7, 1976, 42-

125 Ngoka, N.I.: Development of Low-cost Solar Cookers for Rural Areas of Nigeria. In: [115, 52-56]

126 Nipkow, J.: Sonnenenergie-Geräte für Haushalte in Entwicklungsländern. St. Gallen (Lateinamerikanisches Institut der Hochschule für Wirtschafts- und Sozialwissenschaften) 1977

127 N.N.: Domestic Energy Saving Project: Project Description. Peshawar 1986

128 Nohlen, D. (Hrsg.): Lexikon Dritte Welt. Baden-Baden 1980

129 Nohlen, D., Nuscheler, W. (Hrsg.): Handbuch Dritte Welt. Hamburg 1982

130 Oehler, U.: ULOG Solar Cooker/Oven, Self-Construction Directions. In: [145]

131 Oppen, M. von: The Sun Basket. In: Appropriate Technology, 4, 3, 1977, 8-

132 Osman, M.G.: Solar-Energy for Cooking in Egypt. In: [172, 1768-1795]

133 Pakistan Federal Bureau of Statistics: Household Expenditure Survey 1979. Islamabad 1983

134 Pala, A.O. et al.: The Women's Group Programme in the S.R.D.P. Nairobi 1975

135 Pande, P.C., Thanvi, K.P.: Design and Development of a Solar Cooker for Maximum Energy Capture in Stationary Mode. In: Energy Conversion and Management, 27, 1, 1987, 117-120

136 Parikh, R., Parikh, M.: Design of a Flat-Plate Solar Cooker for Rural Application. Agricultural Tools Research Centre. Bardoli, Gujarat 1978

137 Pheline, J.: Solar Energy (Direct Uses, Water Pumping Irrigation, Greenhouse Heating, Product Drying and Cooking, Environmental Impacts). In: Natural Resources and the Environment Series, 6, 1981, 57-96

138 Philip, S.K., Makwana, H.M.: Final Report of Testing of Stainless Steel and Nonsticking Vessels as Cooking Containers in Solar Cookers. Jyoti Solar Energy Institute. Gujarat Energy Development Agency. Vadodara 1985

139 Philip, S.K., Makwana, H.M.: Final Report on Servicing of Solar Cookers and Developing Public Awareness for Use of Solar Cookers. Vadodara 1985

140 Pinon, R.: About Solar Cookers. In: Passive Solar Journal 2, 2, 1983, 133-

141 Planning and Management Consultancy: A Preinvestment Study for Solar Cooking Box. Khartoum 1986

142 Prakash, J., Sabberwal, S.P., Mishra, R.S., Kishan, H., Bhatia, D.P.: A Rural Solar Cooker. In: [173, 375-382]

143 Proceedings of the Workshop "Transparent Insulation Materials for Passive Solar Energy Utilization", 27th - 28th November, 1986. Freiburg 1986

144 Project Consult (Hrsg.): Arbeitstagung Herdverbreitung. Bericht, Thesen und Ergebnisse. Bad Orb, 1.-6.6.1986. Frankfurt (GTZ/GATE) 1986

145 Ranganath, T.R., Ramachandra, A.: Box Type Solar Cooker with Honeycomb Cover. In: Sunworld 10, 2, 1986, 58-60

146 Salariya, K.S.: Solar Cooking Appliances. In: National Solar Energy Convention of Solar Energy Soc Bhavnagar, India, Dec 20-22, 1978

147 Schäfer, G.: Vorschlag für die Fortsetzung des Domestic Energy Saving Project in Pakistan im Zeitraum 1987/88. Frankfurt (PC) 1986

148 Schwartz, S.: Ökonomie des Hungers - Konsummuster und Vermarktungsverhalten nomadischer Viehhalter Nordkenias. Berlin 1986

149 Schwember, H.E. et al.: Domestic Energy in Sub-Saharan Africa. London (FRIDA) 1980

150 Sepp, C.: Brennholzsparende Herde in afghanischen Flüchtlingslagern in Pakistan. Eschborn (GTZ/GATE) 1984

151 Sepp, C.: Indo-German Integrated Farm Forestry Project. Dhauladhar Palampur, Indien - Verbreitung brennholzsparender Einfachherde (Dhauladhar-Chulha). Eschborn (GTZ/GATE) 1984

152 Sharma, M.P.: Solar Cookers Help in Saving Fuel. In: Non-Conventional Energy Development Agency, India. In: Energy Management-India, 8, 2, 1984, 131-

153 Sid-Ahmed, M.: Solar Cookers Program in Sudan. Wad Medani o.J.

154 Siddiq, A.K., Rabbani, K.S.: Solar Energy Research in Bangladesh. In: ISES/Solar Energy Society of Canada Energex 8th Conference Regina, Canada, Aug 23-29, 1982, 1165-

155 Singhal, A.K., Philip, S.K., Makwana, H.M.: Performance Evaluation of Box Type Solar Cookers, US-India Workshop on Solar Energy. Univ. of Roorkee. India 1985

156 Singhal, A.K., Thierstein, G.E.: The Use of Solar Energy in Post-Harvest Technology. In: [117, 142-159]

157 Smil, V.: Intermediate Energy Technology in China. In: Bulletin Atomic Scientists, 33, 2, 1977, 25-

158 Splitter, E.: Ausbildungsmaßnahmen für Frauen und Mädchen in Bamako. Eschborn (GTZ) 1982

159 Stambolis, C.: Solar Energy Utilization in Developing Countries. In: Ambient Energy, 1, 4, 1980, 219-

160 Statistisches Bundesamt: Länderbericht Indien 1986. Wiesbaden 1986

161 Statistisches Bundesamt: Länderbericht Kenia 1985. Wiesbaden 1985

162 Statistisches Bundesamt: Länderbericht Mali 1986. Wiesbaden 1986

163 Statistisches Bundesamt: Länderbericht Pakistan 1985. Wiesbaden 1985

164 Szokolay, S.V. (ed.): Solar World Congress. Oxford (Pergamon) 1984

165 Swet, C.J.: Thermochemical Storage for Solar Heating and Cooling. In: Sun at Work in Britain, 10, 1980, 20-

166 Thalhammer, T., Prast, G., Langerhorst, J., Floor, W.: Development Cooperation and Solar Energy. In: International Journal Energy Research, 2, 3, 1978, 211-

167 Thomas, J.P., Tully, N.: Field Trials with a Solar Cooker in Rural Southern-Africa. In: [67, 3258-3261]

168 Tully, N., Lawrence, A.G.: The Use of Solar-Energy for Cooking. In: South African Mechanical Engineer, 30, 4, 1980, 118-125

169 United Nations Educational, Scientific and Cultural Organization and World Meteorological Organization (eds.): Solar Energy. Paris 1976

170 United Nations Industrial Development Organization: Technology for Solar Energy Utilization. Vienna (UN) 1978

171 Vaishya, J.S., Tripathi, T.C., Singh, D., Bhawalkar, R.H., Hegde, M.S.: A Hot Box Solar Cooker - Performance Analysis and Testing. In: Energy Conversion and Management, 25, 3, 1985, 373-379

172 Veziroglu, T.N. (ed.): Solar Energy: International Progress. New York (Pergamon) 1980

173 Veziroglu, T.N. (ed.): Renewable Energy Sources: International Progress. Amsterdam (Elsevier) 1984

174 Veziroglu, T.N. (ed.): Alternative Energy Sources VII. Solar Energy. New York (Hemisphere) 1987

175 Veziroglu, T.N. (ed.): Alternative Energy Sources VII. Energy Economics and Planning. New York (Hemisphere) 1987

176 Volunteers in Technical Assistance (VITA) (ed.): Evaluation of Solar Cookers. Mt.Rainier. Maryland 1962

177 Western, Ssemakula: Demand and Supply Futures for Firewood in Kenya. Nairobi 1979

178 Wing, P.: Solar Ovens - Do they Work? In: Arizona Solar Energy Commission, Solar Utilization News, 9, 3, 1984, 8-
179 Winter, F.D., Cox, M. (eds.): Sun. Mankind's Future Source of Energy. New York (Pergamon) 1978
180 Mullick, S.C., Kandpal, T.C., Saxena, A.K.: Thermal Test Procedure for Box Type Solar Cookers. Solar Energy 39, 4, 1987, 353-360
181 Ghougassian, B.: Mectat's Experience in Cooking with Solar Ovens. In: Al-Hamdani, N.I. et al. (eds.): Third Arab Int. Solar Energy Conf. Baghdad 21.-24.2.1988. Proceedings, Baghdad 1988
182 Alward, R. (ed.): Solar Cookers Manual. Technical Report T 138. Brace Res. Inst., St.Anne de Bellevue, Quebec, Canada 1982
183 Bernard, R.: Le soleil à votre table. Découvrez la cuisine solaire. Lyon 1987
184 Magney, G.: Providing solar cookers to Afghan refugees. Proc. ISES Solar World Congr. Hamburg 1987
185 Oehler, U., Oehler, L.: Pakistan-Reise 19.11.-24.12.1986. Basel (unpublished)
186 Energy Res. Inst., Henan Academy of Sciences (eds.): EEB Solar Cooker. Prospectus, Zhengzhou, China (without year)
187 Gangji, Zhu: Recent experience on solar energy utilization in China. Institute of Energy Research, Beijing (without year)
188 Chen, X., Yuanchao, G.: The development and research on solar cookers in China. In: Al-Hamdani, N.I. et al. (eds.): 3rd Arab Int. Solar Energy Conf. Baghdad 21.-24.2.1988. Proceedings, Baghdad 1988
189 Fasholz, J., Cunow, E., Wellers, H.: Die Nutzung regenerativer Energieträger. Rheinisch-Westfälisches Elektrizitätswerk AG (Hrsg.). 2nd edition, Essen 1987
190 Chatelain, H. & R.: Solar cooking of traditional foods in Western Africa. Sunworld 12, 1, 1988, 20-21
191 Khalifa, A.M.A., Taha, M.M., Akyurt, M.: Design, simulation, and testing of a new concentrating type solar cooker. Solar Energy 38, 2, 1987, 79-88
192 Khalifa, A.M.A., Taha, M.M., Akyurt, M.: An energy thrifty solar cooker - the Mina oven. Int. J. of Solar and Wind Tech. 1, 1984, 81-91

193 Olwi, I.A., Khalifa, A.M.A.: Computer simulation of the solar pressure cooker. Solar Energy 40, 3, 1988, 259-268

194 Henzel, V., Holzinger, J.: Solar cooking with a steam immersion heater. In: Bloss, W., Pfisterer, E. (eds.): Advances in Solar Energy Technology. Proc. Bienn. Congr. ISES, Hamburg 1987, 2673-2677

195 Daub, E., Lessle, J., Wippermann, K., Birkholz, U.: Vergleichende Untersuchungen an Solarkocher-Systemen. Proc. 6th Int. Solar Forum, Hamburg 1988, 951-955

196 Heinzel, V., Holzinger, J.: Solar-Dampf-Tauchsieder. Kochen auf andere Art. Sonnenenergie 14, 1, 1989, 26-27

Appendix 1: Solar Cooking Devices

Preliminary notes

Various solar cookers that were devised in developing and industrialized countries for use in developing countries are described below. The individual illustrated descriptions precede a tabular summary comparison.

Most of the data listed in the table were taken from questionnaires that were filled out by individuals and institutions involved in the development and dissemination of solar cookers. A total of 92 individuals and institutions in 31 different countries were written to. Of the 46 replies received, 21 yielded data that could be included in the comparative survey included at the end of this appendix. The various individual solar cooker descriptions contain more such data. It should be noted that practically all such data is more or less subjective, i.e. not free of personal opinion. Some devices are described individually but not listed in the table, which may be taken as an indication of the questionnaire not having been answered properly, so that all or most of the data needed for tabular comparison were missing.

Very little empirical data is available on longevity, so that the information provided is estimative, and the performance data are also influenced by subjective opinions, if not to say extenuation. Even the prices cannot be taken at face value unless a certain minimum quantity of units produced is guaranteed. Frequently, rash conclusions concerning the low cost of future mass-produced devices are drawn on the basis of prototype production.

Questions regarding the cookers' stability and ease of cleaning almost invariably drew such positive answers that the results offered little in the sense of differentiation. Consequently, despite the practical importance of stability and cleanability, those two factors were omitted from the table. Some information on stability is offered in the individual texts.

The definition of efficiency is a critical aspect (cf. sections 2.2 and 3.9). Some respondents went so far as to openly refuse to answer that question. The efficiency of solar cooking boxes and reflector cookers can be described as the ratio between the available heat and the incident energy. To determine the efficiency, the available heat is measured by heating a certain amount of water from a certain initial temperature up to the boiling point. The incident energy is the measured global radiation striking the aperture plane of the device during that same period of time.

But for heat-accumulating solar cookers, only the partial efficiency of the energy-collecting components (collector and heat storage) are directly measurable and, hence, known.

Conspiciously unfounded and/or unrealistic data are shown with question marks and/or in parentheses. To the extent available, more realistic data were inserted.

For the second edition, which was updated and translated to English in 1988/89 some more types of solar cookers have been added both to the descriptive part and to the synoptic comparative survey table.

Box-type solar cookers

Designation:	ATDO Reflector Box Type Solar Cooker
Type:	Solar cooking box with reflecting lid
Designed/produced by:	Appropriate Technology Development Organization (ATDO) Government of Pakistan, Karachi
Promoted/ sponsored by:	All Pakistan Women's Association 67 b, Garden Rd. Karachi 3, Pakistan

ATDO works with several different kinds of box-type solar cookers. They are all larger than a "normal" Indian solar cooking box but designed to hold about the same amount of food. Some (proto)types open at the side (in order to reduce heat losses when opened); the standard version has a tilting double-pane glass cover.

The activities are new; about 20 cookers have been produced since 1986.

Figure 18: ATDO Reflector Box Type Solar Cooker

Designation:	ATRC Solar Cooker
Type:	Solar cooking box with reflecting lid
Invented/ developed by:	Agricultural Tools Research Centre Suruchi Campus, P.O.Box 4 Bardoli - 394 601, India
Manufactured/ marketed by:	Yantra Vidyalaya, P.O.Box 4 Bardoli - 394 601, India

The ATRC solar cooker is a solar cooking box with a reflecting lid. It differs from normal Indian box-type solar cookers mainly by reason of its rectangular shape and stable construction as a robust, all-aluminum-clad wooden box, which also makes it heavy.

While a quadratic shape is optimal for minimizing heat losses through joints and the sides, a rectangular shape has two other advantages:

- it reduces the lateral stress on the hinges when the lid is lifted or lowered unevenly (sometimes problematic for square-bottomed fiber glass cooking boxes), and
- it reduces the amount of radiation lost by not keeping the cooking box accurately (azimuthally) aligned with the sun and/or by neglecting to track the sun at adequately frequent intervals.

At 5 years, the maker's service-life prognosis is quite modest; the device is certainly just as robust, if not more so, than other Indian-made cooking boxes. The cooker comes with a simple, revolving, three-legged stand that puts the unit at a normal working height and facilitates solar tracking.

An appreciable number of ATRC solar cookers have been sold at a quoted price of US$ 60 (converted); however, the major share of the sold units was government subsidized (50 %).

In addition to the solar cooker described here, ATRC also has do-it-yourself instructions for building a cheap, stationary solar cooker. Anchored firmly in the ground, it consists mainly of a double-pane glass cover on a black aluminum receiver surrounded by natural insulating materials (straw or the like) and embedded in the ground. However, the type in question has made no appreciable progress toward dissemination.

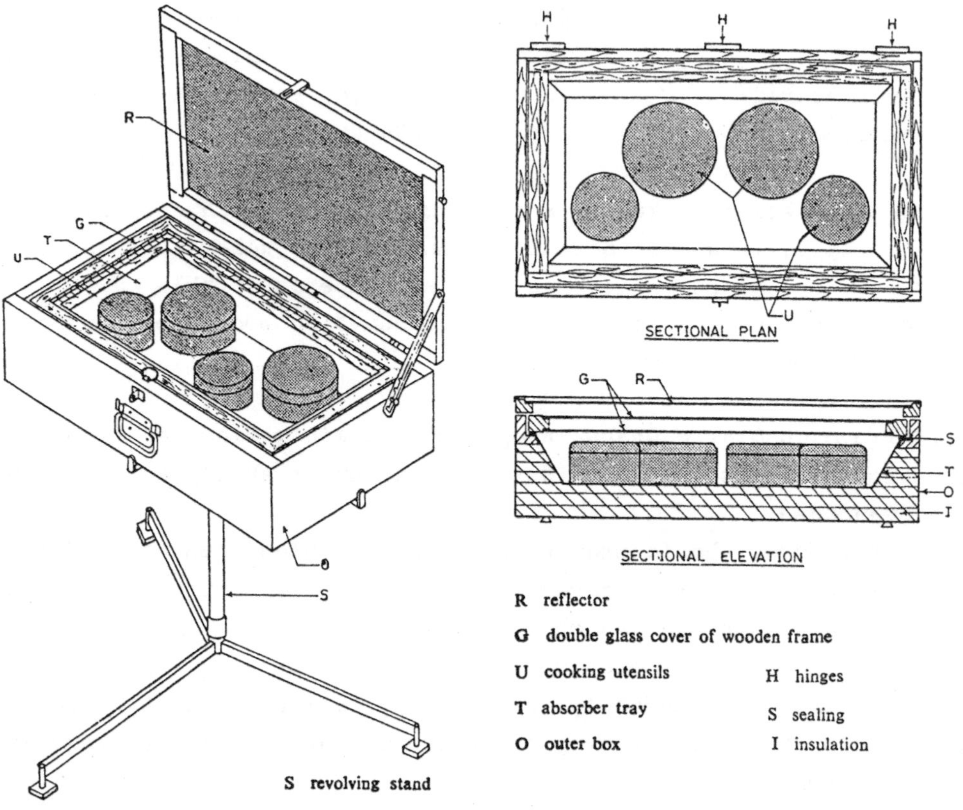

<u>Figure 19:</u> ATRC Solar Cooker on a revolving stand

Designation: Cocina Solar

Type: Box-type solar cooker with reflecting lid

Developed by: Grupo de Investigaciones Agrarias
Academia de Humanismo Cristiano
Casilla 6122 Correo 22
Santiago, Chile

This solar cooking box has an unusually small aperture (28 x 38 cm^2 + mirror of the same size). It is of simple construction and easy to build by following the well-written instructions.

The inside and outside boxes are made completely of wood. The (uninsulated) wooden lid has an aperture-size mirror mounted on the inside. The sides of the inner box reflect radiation, but the black bottom absorbs it. The cover is a single pane of glass.

No information is available on the cooker's thermal output. Its small aperture, its relatively small mirror (which requires constant, very accurate adjustment to reflect fully into the box) and the single-pane cover seem to indicate relatively poor performance. However, the device is promising and capable of improvement.

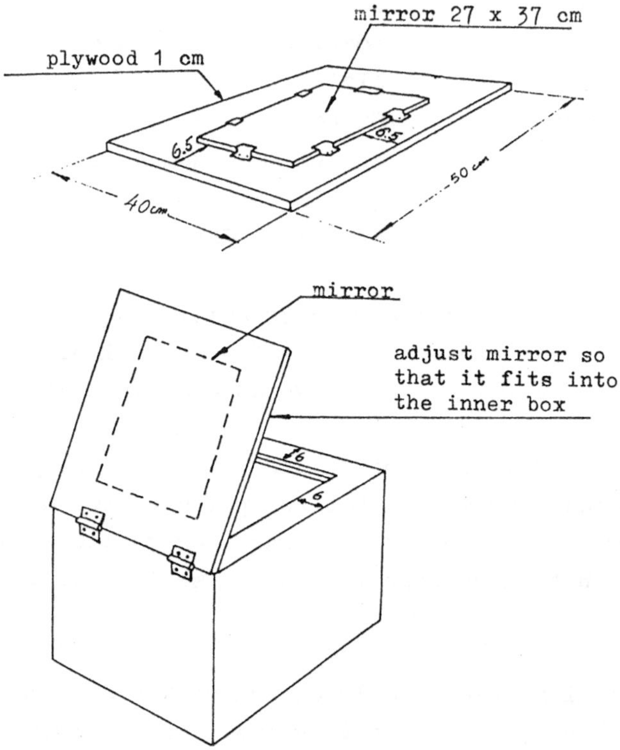

Figure 20: **Cocina Solar**

Designation:	Dhauladhar Solar Cooker
Type:	Stationary adobe cooking box
Developed/ sponsored by:	Mrs. D. Contractor The Mirage V.P.O. Andreta Palampur Tehsil Kangra District H.P. 176-103 India

The Dhauladhar Solar Cooker is a table-level clay-brick solar cooking box, with or without an intensifying reflector.

Larger than most customary types of box-type solar cookers (outside dimensions: 170x75x75 cm; inside dimensions: 86x56x17 cm), this one is designed to handle accordingly large quantities of food (6 kg and more) for large families.

The double-pane glass cover is rigidly attached to the adobe structure (no tilting); the cooking space is accessible from the front, i.e. it corresponds roughly to a drawer. The inclined sheet metal walls of the inner box are made of tin on rice-rusk insulation. Thanks to its modular construction, the individual parts of the inner box can be extracted through the drawer opening without disturbing the glass-cover seal. In practice, this is sometimes necessary for cleaning, repainting, etc. The supporting masonry can be made of bricks or rocks and mortar, while the actual cooking box should consist of adobe clay, since the latter is a good insulator, thus contributing substantially to the box's good heat retention capacity. The cooker still works when individual clouds move past.

Adobe construction makes good weather protection indispensable. In fact making the "umbrella" is part of the building instructions. Nonetheless, the adobe is still subject to gradual erosion, so the user will have to keep it in good repair.

Dhauladhar solar cookers can be built by the users themselves, but they will need the help of an experienced craftsman, since some parts require accurate workmanship.

For better promotion of the cooker, information papers including photos, sketches, recepes, and self-help construction guidelines are being published.

The building materials cost about US$ 22.-, including two shallow pots, but this price does not cover the cost of labour and/or locally available materials. About 150 such cookers have been built since 1985 (see also chapter 2.3.2).

<u>Figure 21:</u> Dhauladhar Solar Cooker

Designation: Indian Box Type Solar Cooker
Type: Solar cooking box with reflecting lid
Institutions: Consortium on Rural Technology (CORT)
E-350 Nirman Vihar
Delhi 110 092, India;
Gujarat Energy Development Agency (GEDA)
B.N. Chambers, R.C. Dutt Rd.
Vadodara 390 005, India
and others
Marketed by: GEDA (see above) and others

The Indian Box Type Solar Cooker is the subject of a national campaign for introducing solar cookers throughout India.

The design and characteristics of the solar cooker are specified by the government and must be adhered to by all manufacturers. The cooker has been type-tested by various institutes. Manufacturers willing to build the cooker according to the official specifications are admitted to the marketing network by the regional or state agency, e.g. GEDA in the state of Gujarat. CORT (an NGO) and the Department of Non-conventional Energy Sources (DNES), Federal Ministry of Energy, among other organizations, are involved at the national level.

The specifications call for fiber glass-reinforced plastic, aluminum-sheet or wood construction. The basic shape is that of a square suitcase, the aperture dimensions amounting to 50 x 50 cm, the lid being fitted with a glass mirror (no reflecting foil) and the inner glass cover pane consisting of tempered (safety) glass. A simple support holds the cover open for introducing and removing food; a second support is provided for adjusting the reflector to the momentary solar altitude. The bottom of the case is fitted with rollers, and a handle is provided for carrying it. Some versions will accept a padlock to protect the food.

Figure 22: Indian Box Type Solar Cooker

Designation:	Kerr-Cole Solar Box Cooker
Type:	Box-type solar cooker made of corrugated paper
Designed/ marketed by:	Kerr Enterprises, Inc. P.O.Box 27417, Tempe Arizona 85282 0410, USA

The Kerr-Cole Solar Box Cooker is a low-cost cardboard model made in appreciable numbers. It has about the same inside dimensions as an Indian Box Type Solar Cooker, but its cover consists of a single pane of glass. A maximum inside temperature of 130...140°C is achievable.

The lid, also made of corrugated paper, has a reflecting foil sticked onto the inside. The optical quality of the reflector is clearly inferior to that of a glass mirror.

Unfortunately, no performance data or prior production figures were available at the time of this writing.

Figure 23: Kerr-Cole Solar Box Cooker

Designation: MECTAT Solar Ovens

Type: Solar Cooking Boxes of different types

Institution: Middle East Centre for the Transfer of Appropriate Technology (MECTAT) P.O.Box 113, 5474 Beirut, Libanon

The Middle East Centre for the Transfer of Appropriate Technology (MECTAT) has developed 9 different types of box type solar cookers (solar ovens), made from different materials such as mild steel sheet, plywood, pressedwood, pasteboard, cardboard. The boxes have partly horizontal, partly inclined (0...40°) glazed covers, different numbers of booster mirrors (0...4), and the apertures vary from 0.08 to 0.26 m^2. There are low-cost models made from recycled materials at approx. US$ 5.00, as well as metal sheet boxes with stainless steel reflectors at the price of US$ 95.00 /181/.

As of this writing, MECTAT is carrying out research on heat storage in pressurized water vessels inside the solar oven in order to make cooking after sunset possible.

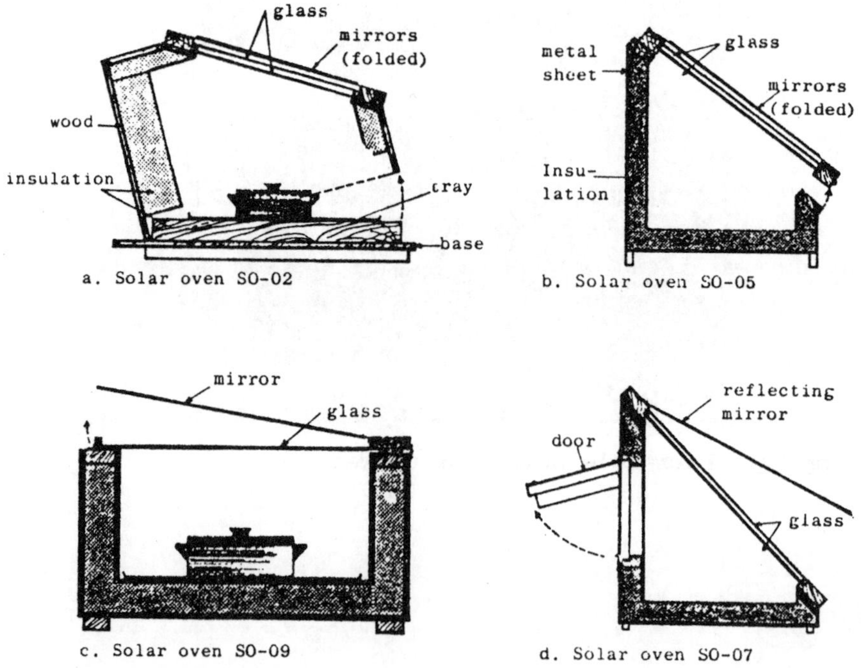

Figure 24: Sectional views of MECTAT's four solar oven designs /181/

Designation:	Mina Solar Pressure Cooker
Type:	Cooking box with booster mirrors and integrated steam pot
Developed by:	A.M. Khalifa, M.M. Taha, M. Akyurt Mech. Eng. Dept., College of Engineering King Abdulaziz University P.O.Box 9027, Jeddah 21413, Saudi Arabia

The Mina Solar Pressure Cooker consists of an air-tight pressure cooking pot with attached heat-conducting metal absorber wings, integrated in an insulated box with a glass or plastic foil cover and a set of 8 booster mirrors in a "funnel" arrangement.

Incoming solar radiation is concentrated by a factor of approx. 3.5 by the booster mirrors, passes through the glass or plastic cover and hits the absorber wings and the pot. Energy is transferred by heat conduction through the wings to the pot walls. The pressure pot allows temperatures over 100°C nearly without evaporation losses, thus enhancing heat conduction within the food and accelerating the cooling process (see ref. 192).

Theoretical studies have shown that the air and vapour tight vessel yields a considerable improvement of efficiency. Smaller cookers of this type perform better than large ones due to their smaller thermal mass. A bigger pot diameter relative to the overall absorber width leads to a higher effective heating power (see ref. 193). Studies aré going on.

Adjusting the cooker to the sun makes inclination of the cooking vessel necessary. This might be acceptable as long as the lid is really air and water tight. Stirring during the cooking process is impossible. Pot handling (taking, pooring out the food, cleaning the pot) is difficult because of the attached absorber wings (the wings must be welded to the pot walls to assure good heat transfer, and cannot be taken away).

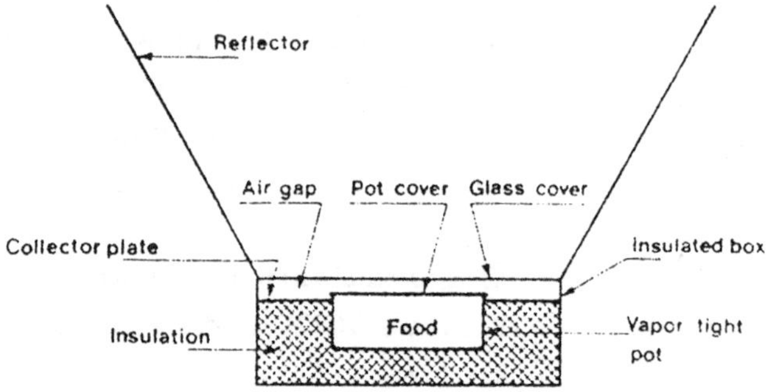

Main components of the solar pressure cooker.

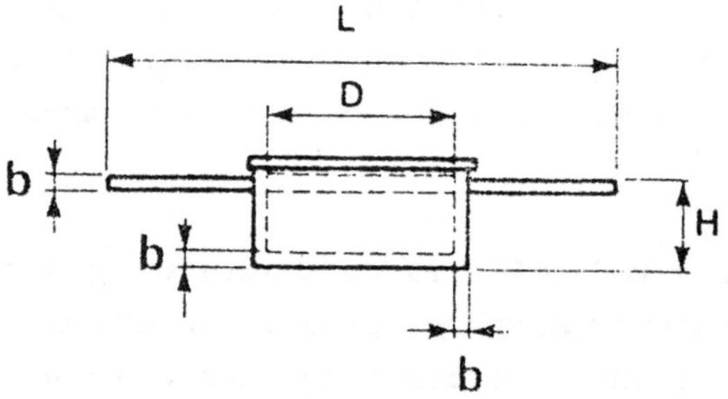

Front view of the integrated pot and collector plate.

<u>Figure 24 a:</u> Mina Solar Pressure Cooker (Saudi Arabia)

Designation:	Orangi Cooking Box
Type:	Box type solar cooker with reflecting lid
Institution:	Orangi Pilot Project, Orangitown, Karachi, Pakistan

An initiative aimed at promoting the construction and distribution of solar cooking boxes emerged in Orangitown (Karachi) in 1985 and 1986 in connection with the Orangi Pilot Project. The cookers in question are of the conventional rectangular type with a reflecting lid. Their main dimensions are 71 x 43 x 20 cm, and the empty unit weighs 20 kg. The somewhat modest maximum temperature of 114°C is related to an equally modest efficiency rating of roughly 25 %. The technical details responsible for the meager performance could not be ascertained.

The production/selling price of US$ 30 would certainly not be too high, if the cooker's performance were commensurate. However, the maximum achievable temperature and performance data are taken as indications of excessively long cooking times, even for standard dishes, so that even small clouds would condemn to failure any attempt to use the box for cooking. In addition, the heavy weight of 20 kg makes it too heavy for a housewife to carry around.

After an initial lot of 30 such cookers was produced, a lack of acceptance led to discontinuation of solar-cooker activities within the Orangi Project. It may be safely assumed that the technical inadequacy of the Orangi cooker - in combination with other socio-eco-cultural factors - prevented it from finding acceptance, so that the project had to be scrapped.

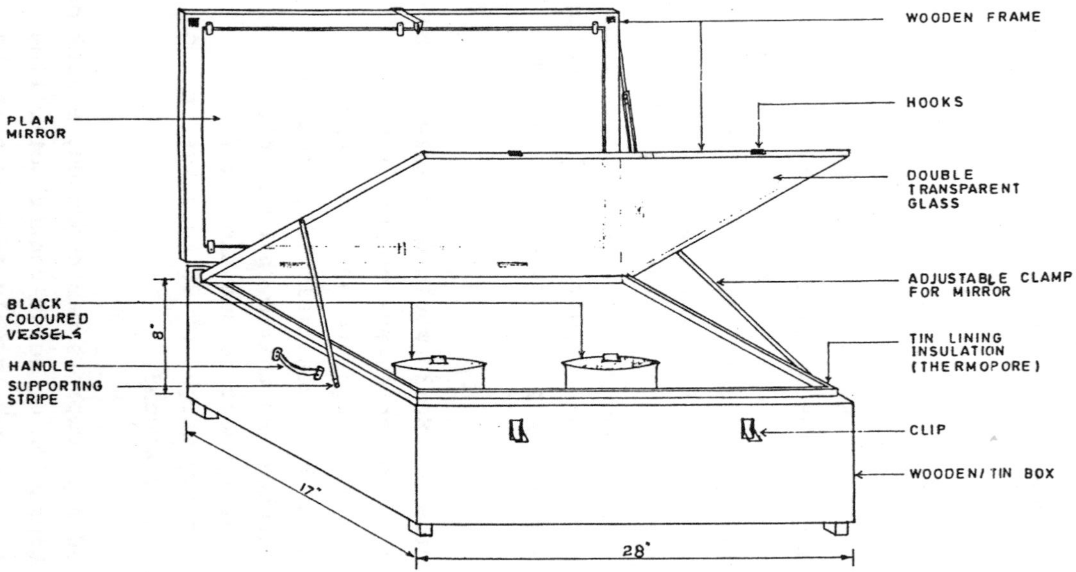

Figure 25: Orangi Solar Cooking Box

Designation: RERI-SEP Solar Cooking Box

Type: Box type solar cooker with reflecting lid

Institutions: Renewable Energy Research Institute (RERI)
Energy Research Council (ERC)
P.O.Box 4032, Khartoum Centre
Special Energy Programme (GTZ)
P.O.Box 8192, Khartoum, Sudan

The solar cooking box being promoted via the Special Energy Programme in the Sudan differs little from other solar cooking boxes in its basic design and construction.

Its basic features are the following:

- the box is rectangular and larger than, say, the Indian type
- it will hold more food
- it is nonetheless handy to use and easy to build
- it cannot be carried like a suitcase
- and is very good at cooking a national dish (foul, a dish based on beans)

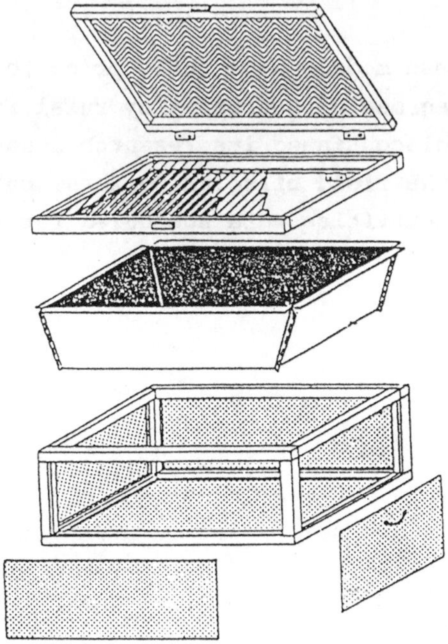

Figure 26: RERI-SEP Solar Cooking Box

Designation:	RIIC Solar Oven
Type:	Inclined box type solar cooker with 4 mirrors
Developed/ tested by:	Rural Industries Innovations Centre (RIIC) Private Bag 11 Kanye, Botswana

The RIIC Solar Oven is a small box type solar cooker with an aperture measuring only 0.1 m^2 in area and an inclined cover with 4 roughly 30 x 30 m^3 mirrors that help concentrate the incident solar radiation (so-called modified TELKES type).

The intensifying mirrors effectively increase the aperture size by a factor of 3. Test reports indicate that this arrangement makes a decisive contribution toward achieving higher temperatures. Consequently, it is reported that the comparatively small receiver can handle a large amount of food (6 kg). The maximum temperature is situated at about 175°C. Nonetheless, the box is reportedly not able to heat water efficiently; in fact, the efficiency amounts to a mere 10 %. A purchase price of US$ 40 is stated (1981).

About 30 of these models were built prior to 1981 and tested - with various degrees of success - by rural families.
In 1981, RIIC discontinued its research & development activities in the field of solar cookers, but, as of this writing, such activities were scheduled for renewal in 1986/87.

Designation: SERVE Solar Oven

Type: Inclined solar cooking box with reflecting lid

Dissemination by: SERVE, P.O.Box 477
Peshawar, Pakistan

The SERVE Solar Oven is a rectangular solar cooking box with double glass cover, a reflecting lid, and approximately the double size of the Indian solar cooking box (width about 1 m).

In the 1987 model, several improvements have been realized, such as 20° inclination of the glass cover, Mylar foil reflector instead of glass mirror, fiber glass casing instead of sheet metal box, etc. These measures reduced the weight and raised the efficiency of the oven.

As well, the lower glass of the cover was cut into two pieces, then sealed again with silicone to make the glass less susceptible to thermal tension. Thus, glass breakage was reduced from 15 % to less than 5 % /184/.

More than 4000 ovens of previous design types (square shape, horizontal cover, etc.) have been produced by 1987. For the improved 1987 type, a production of 3700 items p.a. has been planned.

Figure 27: SERVE Solar Oven, improved 1987 model

Designation: Suryamuklu Box Type Cooker
Type: Box type solar cooker with reflecting lid
Produced/marketed by: Centre of Energy Studies IIT Delhi

The Suryamuklu Cooker is a technically improved version of the Indian Box Type Solar Cooker and costs only about half as much to manufacture. It is substantially smaller, easier to use, and reportedly handles the same amount of food as an Indian Box Type Solar Cooker - even during periods of weaker insulation (or in less time). However, such information stands in contradiction to the box's very low maximum achievable temperature of 100°C. No reliable performance data were available.

Production has either not yet started or only did so recently. No illustration/photo and no details on efficiency-enhancing measures were available as of this writing.

Designation:	**22° Solar Cooker**
Type:	**Lightweight box type solar cooker**
Developed/ produced by:	**I. de Klerk** **Harare Polytechnic** **Box 8074** **Harare, Zimbabwe**

The 22° Solar Cooker is a do-it-yourself cooking box made of glued cardboard. The name derives from the top's angle of inclination, which roughly corresponds to the geographic latitude of Harare (18°S).

According to the instructions, the user is free to size the box as desired. About 21 of these solar cookers have been built since 1982.

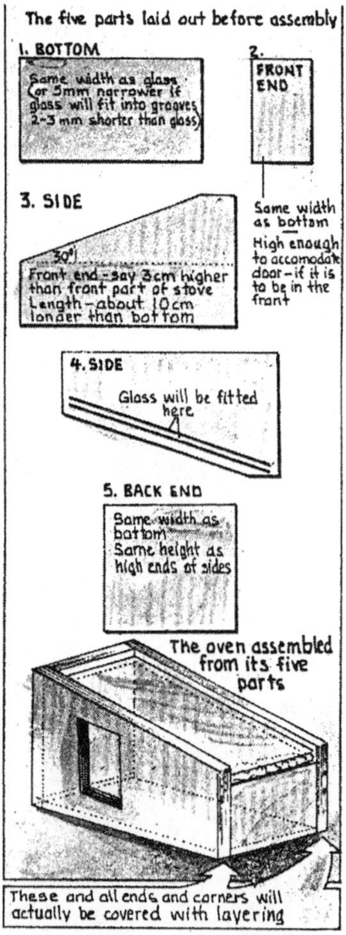

<u>Figure 28:</u> 22° Solar Cooker

Designation:	ULOG-Tropical Solar Cooker, Model 85
Type:	Box type solar cooker with reflecting lid
Invented/ developed by:	U. Oehler Morgartenring 18 CH-4054 Basel
Institution/ marketed by:	U. Oehler

The ULOG-Tropical Solar Cooker, model 85, is a stable, robust, wooden solar cooking box designed for self construction or commercial fabrication by simple means.

The box is quadratic, covered with two panes of glass and a wooden lid with a reflecting foil on the inside to serve as an intensifier for cooking.

Figure 29: ULOG-Tropical Solar Cooker, Model 85

The sides of the inner box are slanted. The floor is painted flat black, and the walls are silvery reflecting. Used sheets of aluminum offset-printing foil are used for lining the inner box. In fact, the cooker design is based on the dimensions of those foils. The stiff aluminum foil is bent such as to produce a thoroughly vapor proof box (important, because the insulation between the inner and outer boxes must be kept dry).

The wooden outer box is built for stability, not for light weight. By employing a few tricks of the trade, the assembly procedure was kept relatively simple without sacrificing any of the precision that is necessary to keep the cooker from losing heat through cracks and joints.

Well-illustrated and equally well-written assembly instructions are available in English, French and German (Spanish and Portuguese in preparation). The self-construction directions are included here as Appendix 2.

One drawback of the Model 85 Tropical Solar Cooker is its somewhat bulky shape due to stable construction. It has two handles but cannot be carried like a suitcase.

Its predecessor, the ULOG Tropical Solar Cooker, Model 80 (rectangular) could be carried like a suitcase, but it also had less space inside (for not more than 3 cooking vessels).

About 450 ULOG Tropical Solar Cookers (both models) have been built to date, 90 of these in Sudan, 60 in Senegal, 100 in Turkey, etc.

<u>Figure 30:</u> ULOG Tropical Solar Cooker, Model 85

Designation:	ULOG European Solar Cooker
Type:	Box type solar cooker with inclined cover
Invented/ developed by:	U. Oehler, Morgartenring 18 CH-4054 Basel
Institution/ marketed by:	U. Oehler

The ULOG European Solar Cooker is very similar to the ULOG Tropical Solar Cooker model 85 in design and construction, except that it has an inclined cover that allows the capture of enough insulation for cooking in moderate climates (where the sun has a lower altitude).

The ULOG cooking box has also been successfully tested in combination with a heat store consisting of 4 kg dark-colored stone. For the stones to accumulate enough heat for cooking, the device must be set up early in the morning. By the time the food is loaded into the box, the stored heat provides a relatively high starting power. It also serves well as a source of bottom heat for baking.

Some 1600 European Solar Cookers have been built (from kits and in home construction), most of them in Switzerland. A set of instructions can be obtained from U. Oehler.

Figure 31: **ULOG European Solar Cooker**

Designation: Four-mirror Cooking Box

Type: Box type solar cooker with inclined aperture and four intensifying mirrors

Developed by: Gabriel Rodriguez Jaque
Sección Fisica de la Construcción en IDIEM
Universidad de Chile, Plaza Ercilla 847-2°
Casilla 1420 Santiago, Chile

This solar cooking box measures 0.9 x 0.9 m^2 at the base and has a 0.49 m^2 aperture set at an angle of 30°. Four mirrors of equal size are hinged on around the top of the box to provide intensified radiation input.

Performance data: Depending on the season, location and momentary weather conditions, the device can collect up to 2.3 kWh per day, or enough to heat 18 l of water from 15°C (average spring temperature in Chile) to 95°C.

The prototype for this cooker took first prize in a UNESCO-sponsored solar cooker competition in Chile /65/.

Figure 32: Four-mirror Cooking Box

Reflector Cookers

Designation: Type:	Advanced Reflector Cooker for Mali
Invented/ developed by:	Arbeitsgemeinschaft für Entwicklungsplanung (AE), Munich
Project-executing agency:	Deutsche Gesellschaft für Technische Zusammenarbeit (GTZ) GmbH, P.O.Box 5180, D-6236 Eschborn

The Advanced Reflector Cooker was developed by AE München /6/ in 1979/80 on the basis of the then latest knowledge concerning solar cookers. The design stage of development was preceded by a thorough global research effort involving detailed testing and evaluation of sundry solar cookers /5/. The advanced reflector cooker was supposed to emerge from knowledge gained on other devices and incorporate/optimize as many positive characteristics as possible.

The basic unit consisted of a short-focusing glass-reinforced polyester reflector shell in a swivel mount on a frame made of angle irons and/or square bar steel. The diameter of the reflector was 1.4 m, and the shell had cutouts on three sides, so that

1) the load-bearing frame and, hence, the entire reflector cooker would be much more compact and stable, and
2) the cook could use one of the cutouts to get close enough to the pot for comfortable handling and stirring.

Other improvements over prior-art reflector cookers included:

3) coincident pivotal point, center of gravity and focal point of the reflector shell, thus enabling easy adjustment without the danger of injury caused by the reflector swinging back;
4) the option of turning the shell away and out of the sun to prevent accidents and avoid unnecessary soiling of the reflective surface while manipulating the pot;
5) minimum convection losses, since the pot was protected from the wind by virtue of its position below the shell's rim;

6) the possibility of turning over the swivel-mounted reflector when the device was not in use in order to protect it from damage and dust accumulation;
7) an additional worktop (pot setdown area) provided at normal working height beside the shell;
8) the extreme importance attached to stability, e.g. for stirring;
9) but with the cooker still remaining so light (10-kg reflector, 15-kg frame) and compact that it could be brought into position by one person, meaning that no supplementary adjusting axis, e.g. parallel to the earth's axis, was necessary;
10) the short focal length and the in-reflector position of the pot, yielding uniform heating from all sides at once and, hence, good efficiency (59 % at 770 W/m^2);
11) the short focal length (0.26 m), permitting slight inaccuracies in the shape of the parabola without undue loss of efficiency.

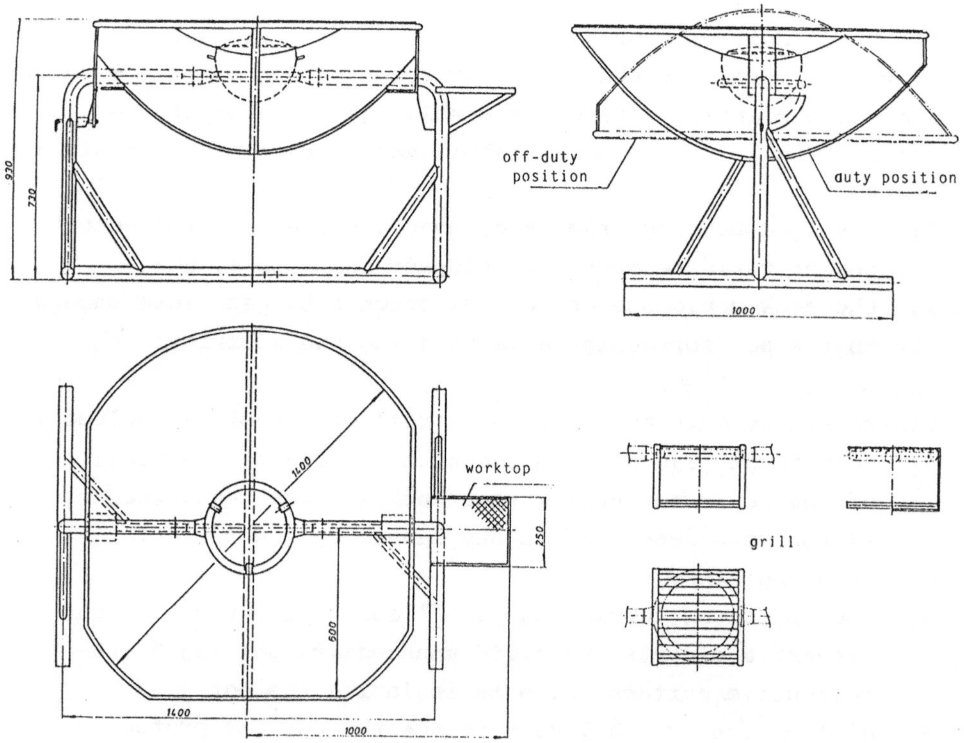

Figure 32a: Advanced Reflector Cooker for Mali, 1980

The cooker's maximum capacity was 10 l of food.

The improved reflector cooker naturally had some of the same drawbacks with which all reflector cookers are afflicted, e.g. no utilization of diffuse radiation, necessity of standing in the sun for cooking, no chance of cooking under cloud cover, necessity of stirring due to (excessively) high temperatures, solar tracking, no attendance-free cooking, no warm food in the evening.

The aluminum-coated PE foil used for lining the inside of the reflector caused problems during prototype development as well as during the subsequent testing phase.

Alternative prototypes were made with wooden supports and papier maché reflector shells, but had to be rejected for lack of stability. Various experiments were conducted (some in the Federal Republic of Germany and some in Mali) with smaller and larger diameters, other kinds of foil lining, etc. Various design modifications were tried out in order to make the device as inexpensive as possible and more amenable to locally available materials.

The cost of producing a prototype in 1980 came to roughly US$ 100, or about as much as a large family in Mali would save on firewood in the course of 1 1/2 years by using a solar cooker.

About 16 such devices were built. Field testing of 10 units in Mali yielded negative results, due in part to organizing problems and in part to the inherent disadvantages of cooking with a solar reflector cooker in general.

In sum, it may be said that, thanks to the aforementioned improvements, the Advanced Reflector Cooker probably was - despite its drawbacks - one of the best-conceived (from a technical standpoint) solar reflector cookers ever built (cf. Sun Basket and Falco S/C).

Designation: Bottom-Heated Concentrator Box

Type: Bottom-heated box-type solar cooker with spiral concentrator

Developed by: A.M.A. Khalifa, M.M.A. Taha, M. Akyurt
Mech. Eng. Dept., College of Engineering
King Abdulaziz University
P.O.Box 9027, Jeddah 21413, Saudi Arabia

The bottom-heated concentrator box is the result of the inventors' efforts to combine efficiency, ease of construction and operation, and simplicity of box type solar cookers with the suspension principle of the Telkes design and the concentrated energy flow and high power of concentrator s.c.s.

An insulated cooking box with glass bottom and side walls is suspended by a metal rod frame in the focal spot on top of a spiral metal sheet reflector of 1.13 m^2 aperture area. The pot is heated from the bottom and not - as in most of the other solar cooking boxes, from the lid. This makes heating more effective and cooking times shorter. High energy flux and good box insulation lead easily to temperatures about 180°C. This enhances the possibilities of food processing: Grilling and roasting are possible.

A boiling test of 6 kg of water resulted in an efficiency of approx. 18.5%. The heating process does almost not depend on wind velocity as it does for most reflector cookers. Sun tracking of the concentrator does not affect the horizontal position of the cooking pot; food does not spill over.

Handling of the cooker is not easy, stirring the food during the cooking process is hardly possible, though it is necessary for many types of food when cooked in a concentrator (turn the reflector away, open the swinging box, open the pot, stir in the swinging device...). Research and development are going on (see ref. 191).

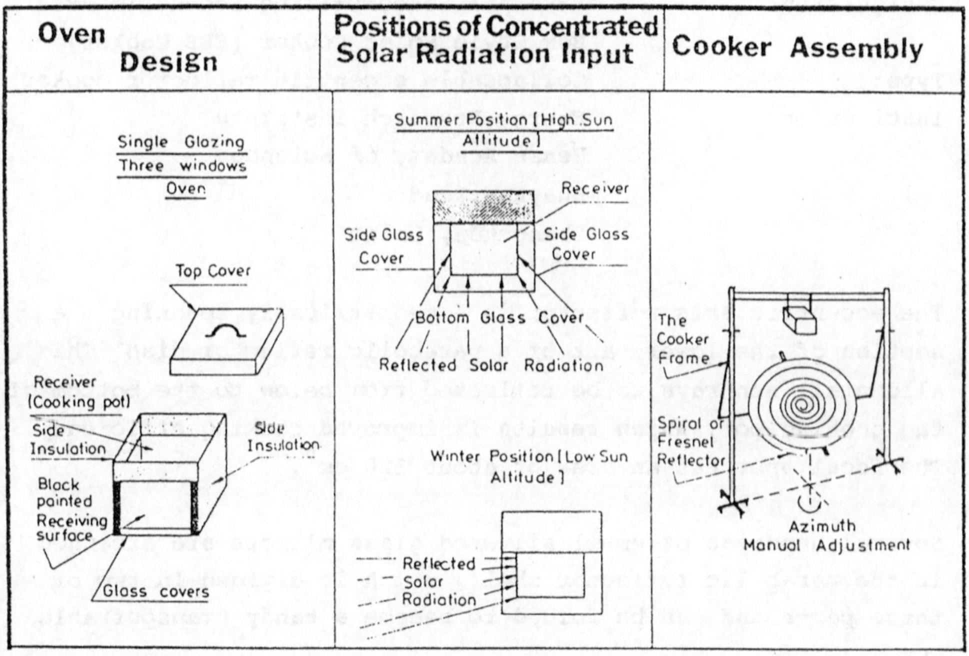

<u>Figure 32 b:</u> Bottom-heated concentrator box (from /191/)

Designation: External Concentrating Eccentric Axis Box Style Solar Cooker (EEB Cooker)

Type: Collapsable eccentric reflector cooker

Institution: Energy Research Institute
Henan Academy of Sciences
Huayuan Road
Zhengzhou, China

The eccentric axis reflector is - geometrically speaking - a section of the lower part of a parabolic reflector dish. This allows all sun rays to be reflected from below to the bottom of the cooking pot, which results in improved cooking efficiency. The focal spot has an area of about 120 cm^2.

Several hundreds of small silvered glass mirrors are arranged in the parabolic reflector shell, which is divided in two or three parts and can be folded to become a handy transportable box.

The radiation collecting aperture is about 1 m^2, the effective power about 500 W on sunny days. Steaming, boiling, and frying is possible in the cooker. The working height (cooking pot level) is between 1 m and 1.25 m. It can provide a meal for 3 - 4 persons.

Other external concentrating eccentric axis solar cookers are made from fiberglass reinforced cement. Some have an aluminum foil reflecting surface. Nearly all of the solar cookers in use in China are of the external concentrating eccentric axis type (see figs. 13 and 32 c).

Figure 32 c: External concentrating eccentric axis box-style (EEB) solar cooker, China

Designation:	Falco S/C
Type:	Parabolic reflector cooker
Developed/ produced by:	B.B.K. J.D. Kaller Kleingsenget Haus 111, P.O.Box 11 D-8391 Neureichenau, W-Germany

The Falco S/C is a reflector cooker with a medium focal length. It is conspicuous by virtue of its simple construction.

The mirror is adjustably mounted in a simple frame equipped with a simple chain mechanism to keep it in line with the sun's altitude. The axis of rotation itself is stationary, running through the center point of the mirror and including a holder for the pot (or two) at the focal spot.

The metal mirror is an industrial, single-piece product with a diameter of 115 cm. The polished surface has a reflectance of about 80 %. A significant share of the incident radiation is reflected diffusely and therefore unavailable as cooking energy.

The complete unit, including mirror and frame, weighs less than 20 kg. It comes with two shallow flat-bottomed pots (3 l and 4 l) that can be used simultaneously, i.e. while food is cooking in one pot, the other pot can be set on top of the first for slow cooking, warming up or keeping food warm.

Efficiency of between 20 % and 35 % for global radiation intensities above 800 W/m^2 and wind velocities of mostly over 1 m/s were measured at the RERI Institute in Karthoum. The modest efficiency is probably due to diffuse reflection (= energy loss) by the parabolic reflector, in combination with the fact that the pots project beyond the spherical segment and are therefore exposed to the wind. In addition, the absorptance of the pots appears less than optimal.

In quantities of 100 or more, the Falco S/C costs US$ 240 apiece. To the extent possible, regionally dependent "special options" are allowed for at the time of production.

Figure 33: Falco S/C Parabolic Reflector Cooker

Designation:	Fixed-Focus Hybrid Solar Cooker
Type:	Hybrid wood and solar off-axis parabolic reflector cooker
Invented/ developed by:	W. Scheffler c/o German Water Team P.O.Box 19512 Nairobi, Kenya

Fixed-focus solar cookers are composed of one or more off-axis parabolic mirrors and a stand for one or more pots with a secondary mirror underneath which reflects the concentrated radiation coming from the parabolics, to the bottom of the cooking vessel(s).

Furthermore, these cookers are hybrid s.c.s. This means that, whenever there is not sufficient direct solar radiation, the secondary reflector under the cooking pot can be removed, and the pot can be heated by a conventional wood fire.

3 fixed-focus hybrid solar cookers have been established in Kenya till 1988, one family-sized device and two big ones for institutional cooking at schools. The latter are equiped with an additional tank for warm water generation.

The technical data can be taken from the following table:

Nr.	1	2	3
year	1986	1987	1988
location	Mission North Horr	St. Joseph's Girls School Kibwezi	Primary School Kalacha
mirrors, aperture	1 x 1 m^2	2 x 3.8 m^2	4 x 5.6 m^2
heating power (kW)	0.4	1.7	5.6
efficiency (%)	36	28	31
no. of pots	> 2	1	2
total amount of food (l)	5	54	200
warm water tank (l)	-	136	270
firewood saved per day (kg)	-	12	50
investment (US$)	112.-	2250.-	3350.-

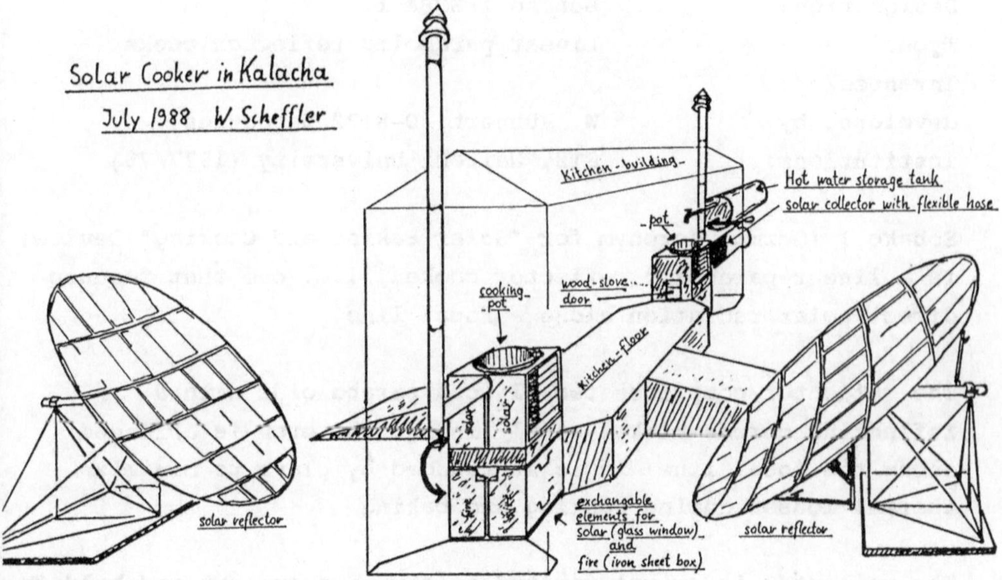

Figure 33 a: **Fixed-focus hybrid solar cooker, Kenya**

Designation: Sobako 1/SOBA 1
Type: Linear parabolic reflector cooker
Invented/
developed by: W. Kühnert, D-8172 Lenggries
Institutions: GTZ, Nairobi University (1977/78)

Sobako 1 (German acronym for "Solar Baking and Cooking" Device) is a linear-parabolic reflector cooker, i.e. one that focuses direct solar radiation along a focal line.

The reflector comprises two plywood paraboloids with a reflecting chrome-nickel steel lining. The oven is arranged along the focal line; it is surrounded by glass to minimize thermal losses during cooking and baking.

The reflector is swivel-mounted on a wooden support and held in place by a counterweight. It has to be realigned with the sun by hand every 15...20 minutes.

The cooker's span width comes to 2.46 m, the width of the reflector/oven is 84 cm. The aperture figures to just over 2 m^2. The inside diameter of the oven is 11.5 cm. The cooking capacity is listed as adequate for preparing a rich European midday meal for four or for baking two 1-kg loaves of whole-grain bread at once.

When not in use, the cooker can be folded up to protect the internals and the reflector from the effects of wind, rain and dust. A cover is also provided.

Considering the unit's cooking/baking capacity (2 kg bread or food) its collecting area of roughly 2 m^2 and span width of nearly 2.5 m appear quite large, and its mechanical design is relatively elaborate (numerous different parts). Additionally, its projecting structure necessitates a broad, heavy, revolving foot.

At 170...180°C, the reported maximum achievable temperature inside the oven is rather modest, considering how large the collecting area is. This corresponds to other reports, according to which the special steel reflectors are quick to dull, after which they are heated by the sun and display poor reflectance (estimated 50 %).

According to the reports (with the exception of the designer's), Sobako 1 was expensive, complicated, unfunctional and awkward to use. In other words, it was inappropriate.

Some of the points drawing criticism were that
- even two pepole had a hard time transporting the device,
- setting it up required so much technical knowledge, carefulness and physical strength that the users were hardpressed to accomplish the task,
- the oven had to be taken apart for cleaning,
- the insulating glasses surrounding the oven were inadequately mounted,
- the mirrors had a blinding glare,
- the cooking vessels intended for use in the oven were prone to rusting,
- the oven was 1 cm too small for standard-size pots,
- even hazy weather (no clouds!) made cooking impossible,
- the cooking capacity was too small for a "normal family" with 4 to 8 children,
- the device had to be made of high-quality waterproof plywood, with no possibility of using alternative materials,
- various parts of the cooker tended to swell up and become distorted due to the effects of heat and humidity,
- the oven was heated unevenly,
- the necessary tracking interval of 10 minutes required someone's constant attention,
- the device hardly ever had enough direct radiation to work efficiently in Nairobi's constantly smoggy atmosphere,
- it was incompatible with local eating and cooking habits, and, finally,
- it "disqualified itself by reason of its exorbitant cost".

Sobako 1 was succeeded by a simplified type called Soba 1. Each of the two prototypes built at the University of Nairobi cost well over US$ 420. Assuming otherwise favorable conditions and production in considerable quantities, a prime cost of US$ 252 was regarded as achievable. Payback periods of 3.5 years based on fuel savings and 1.8 years based on the added advantage of home-baked bread were calculated. Partial improvement of the reflecting liner yielded higher efficiency.

Plans were drawn up for series production and dissemination, and recommendations concerning a subsidization program were elaborated for submittal to the government. A successor type Soba 2 was never built. The high cost of construction, awkward handling (even of the simplified Soba device), the natives' habit of cooking their main meal in the evening, the cooker's dependence on direct radiation and the practically permanent haziness in and around Nairobi were presumably the main reasons why Sobako 1 and Soba 1 never got past the prototype testing stage.

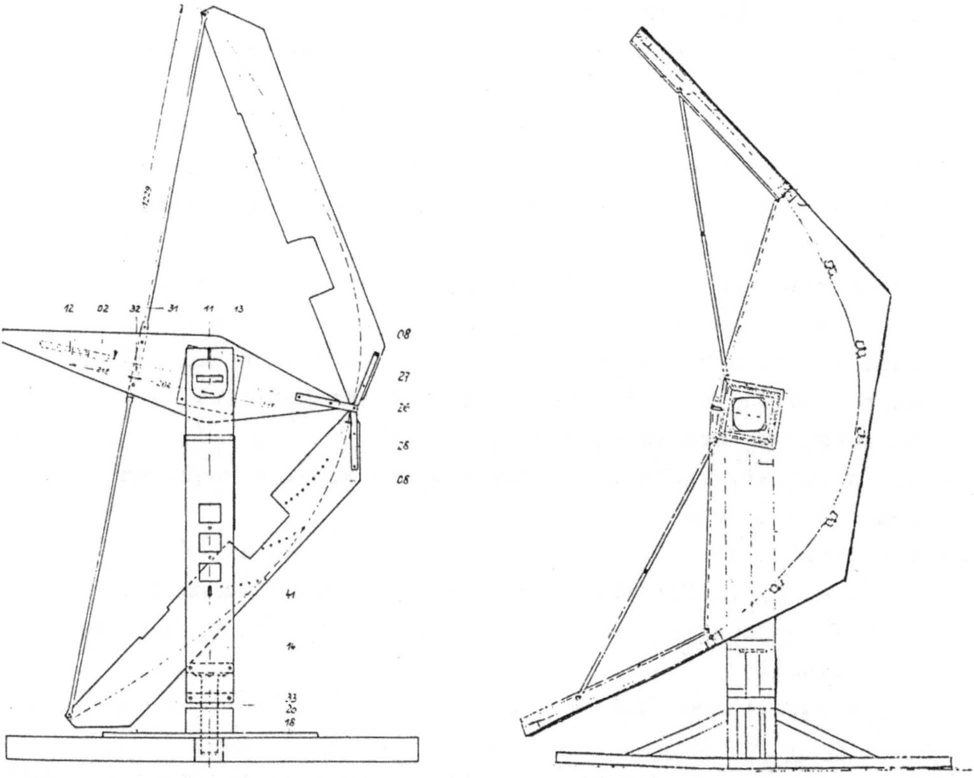

Figure 34: Sobako 1 (at left) and Soba 1 (at right)

Designation:	Sun Basket
Type:	Reflector concentrator with semi-automatic tracking feature
Invented/developed by:	Dr. M. v. Oppen, Steinwaldstr. 25, Hohenheim-Stechfeld, 7000 Stuttgart 70, Fed. Rep. of Germany
Manufactured/marketed by:	Mr. Kiram Chandwalker, Stiletto Engineers Shed No. F5/A, IDA, Kukatpally Hyderabad 500 037, India
Sponsored by:	Sonnenkorb Association Barckhausenstr. 47 D-2120 Lüneburg, Fed. Rep. of Germany

Sun Basket is a large, relatively short-focussing parabolic reflector in which a pot is held in the focal point by an attachment on an iron rod. The parabolic reflector measures about 1.3 m in diameter and is made of glass-reinforced plastic (GRP, fiberglass) over a masoned mold. It is relatively expensive to make and accounts for most of the device's unit price. The reflective liner consists of vapor-deposited aluminum foil, which must be replaced periodically.

Earlier Sun Basket models had parabolic reflectors made of curved strips of bamboo - hence the name. But problems with stability and accurate focussing made it necessary to discontinue that very inexpensive type of construction.

A different Sun Basket model has an ingenious solar tracking mechanism that operates according to the water-clock principle in keeping the reflector almost exactly aligned with the sun. The pot hangs on a sort of jib over or in the bowl of the reflector. While this arrangement is quite elegant, it also has the drawback of having to hold the pot by hand for stirring its contents - and manipulating the "floating" pot in the cooker's focal point poses the danger of injury by burning. That model has also been discontinued.

The reflector is quite large and has no recess that would let the cook get closer to the pot (as the Advanced Mali Cooker has). Having to lean over the basket to stir food, for example, is not only tedious but also hazardous (blinding, burning). In addition, the cook must stand out in the sun.

About 20 such Sun Baskets have been built since 1985. Work is still being invested in making the reflector more reflective and less expensive.

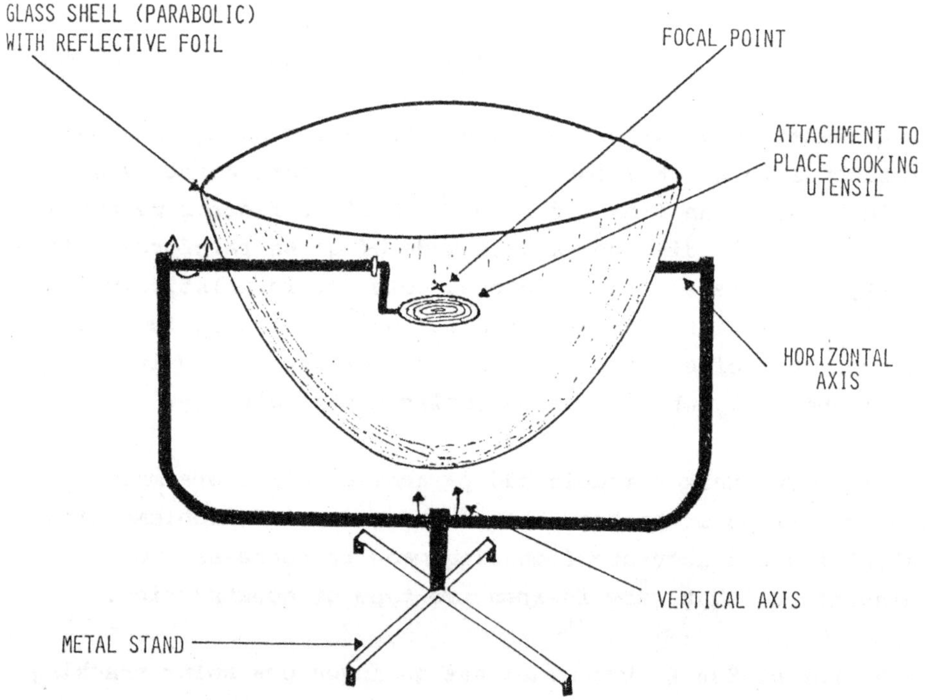

Figure 35: Sun Basket (1985)

Designation:	Sungril
Type:	Collapsable transportable aluminum reflector cooker
Manufacturer:	Interel S.A., Clarastr. 2, CH-4005 Basel, Switzerland

Sungril is an industrially manufactured, light-weight collapsable reflector cooker of 1 m diameter which can easily be transported by hand in a 50 x 36 x 12 cm briefcase box. The pot holder is fixed on a vertical stand in the 15 cm diam. focal spot. It does not move when the reflector is adjusted to track the sun's position. All elements are made from aluminum and free of corrosion.

The light-weight construction is easy in handling and transport, but not very stable when stirring of the food is required or the wind is blowing.

Sungril has been produced for 10 years and is patented in Switzerland and the USA.

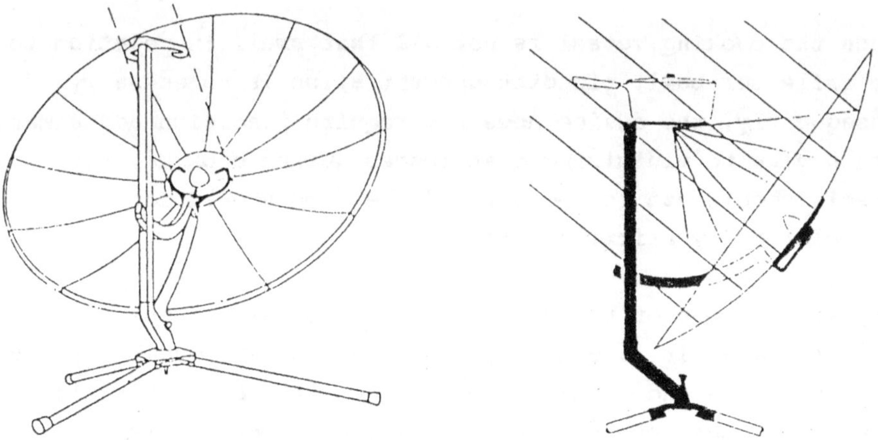

Figure 35 a: Sungril

Designation:	Suryakund
Type:	Small, nonfocussing elliptic reflector cooker with glass cooking vessel
Contact:	CORT (cf. Indian Box Type Solar Cooker) and others

The Suryakund is the simplest known type of reflector cooker. It comprises four parts:

- an asymmetrical reflector bowl made of plastic, with an aperture size of approx. 0.2 m^2,
- an asymmetrical support that will accommodate the reflector bowl in practically any position,
- cooking utensils: two black pots (1-liter and 2-liter capacity) that fit together (nest) and have tight-fitting lids, and
- a glass globe to reduce heat losses.

The food is put in one or both pots. The pot(s) is/are placed in the globe, which completely surrounds the pot(s) and retains heat by virtue of its greenhouse effect. Then, the ensemble is placed in the reflector's concentrated field of solar radiation. The solar rays penetrate the globe to uniformly heat the bottom part of the cylindrical pot.

Since the cooking vessel is not all that small in relation to the reflector bowl, pinpoint concentration is unnecessary. Consequently, the device does not require precision adjustment. With a view to minimizing heat losses during cooking, the vessels should not be opened, and the food should not be manipulated (by stirring, etc.).

Suryakund is one of the simplest solar cookers in the world. Made in India, it is sold in large numbers. Its drawbacks are its small area and small capacity, i.e. it doesn't readily accept more than one pot at a time, and it mostly makes use of direct solar radiation only.

Figure 36: **Suryakund Reflector Cooker**

Designation: Table-type Reflector Cooker
Type: Reflector cooker with its own worktop
Developed by: Roger Bernard, Laboratoire d'Energie Solaire, Université Claude Bernard, 43 Boulevard du 11 Novembre 1918, F-69622 Villeurbanne, Cedex, France

The Table-type Reflector Cooker is a typical reflector cooker with no means of heat accumulation, but which avoids some of the drawbacks of conventional reflector cookers.

The "hearth" is stably integrated into a worktop standing on four legs about 70 cm above the adjustable reflector. This extensively precludes problems concerning pot-mount stability and general handling.

The rectangular parabolic reflector consists of 35 flat mirrors. Its maximum thermal output in bright sunshine peaks below 400 W. That amount of energy generates a temperature well in excess of 200°C at the bottom of an empty pot. For a more detailed description see refs /11/, /12/, /183/.

This cooker has a noteworthy adjusting mechanism: a small positioning mirror reflects onto a focusing plate used for checking the momentary quality of alignment. When the unit is not needed for cooking, it can be put aside with its reflector turned vertical or upside down.

Apart from the aforementioned merits, this cooker also has some of the drawbacks of conventional reflector cookers: the cook has to stand out in the sun, and cooking is impossible without adequate direct radiation, which excludes cloudy days and the morning/evening hours.

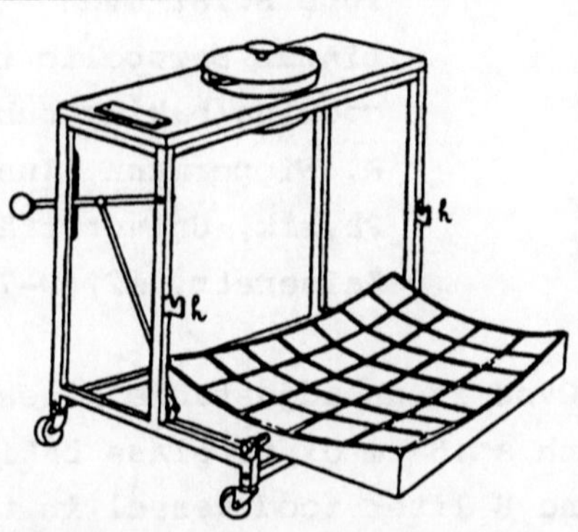

Schematic view of the tropical prototype.

<u>Figure 37:</u> **Table-type Reflector Cooker**

Designation: Tube Solar Oven

Type: Linear parabolic reflector cooker with cooking/baking tube

developed by: K. Wippermann, Inst. f. Angewandte Physik, Universität Karlsruhe, Kaiserstr. 12, D-7500 Karlsruhe, FRG

The Tube Solar Oven is an adjustable linear parabolic 0.6 x 1.0 m^2 reflector with a 15 cm diam. glass baking/cooking tube containing a long 8 liter food vessel in the focal line. At first glance, it has the appearance of a comfortable arm chair.

The main features of this s.c. are:

- The mirror is a one-dimensional parabolic reflector which is easy to manufacture.
- Solar tracking is simple and necessary only every 40 min because of the relatively wide acceptance angle of the tube (15°).
- The glass tube acts as a collector, collects direct, diffuse, and indirect concentrated radiation from all sides and therefore does not need any insulation.
- The parasitic (= unwanted) heat capacity is low.
- The energy flow to the food vessel is high enough to make brown cake and bread, but low enough to make stirring unnecessary.
- Normal cooking pots cannot be used; special cooking vessels are required.

Optimization has lead to a medium-cost low-capacity, high-efficiency construction with 46 % efficiency in the water heating test. Till 1988, one prototype was constructed; research is going on in the directions of heat accumulation and steam generation for sterilization.

<u>Figure 37 a:</u> **Tube Solar oven**

Designation: Valparaiso Reflector Cooker
Type: Parabolic reflector with mirror mosaic
Developed by: Jaime Lopez C., Luis Seguel R., Juan Jerez I., Universidad de Diseno, Av. El Parque s/N. PYA. Ancha, Valparaiso, Chile

This cooker is characterized by extreme simplicity. A well-conceived set of instructions makes it relatively easy to build and use. Building is possible at home (self-made) or by local craftsmen.

The parabolic shape of the shell is achieved on a stationary adobe mold with a reinforced-concrete core. The inside of the parabola is studded with numerous fragments of mirrors. The cooking pot rests on a wire grate situated at the cooker's focal point. The grate hangs on the two arms of a supporting fork such that the pot always maintains a horizontal position. The reflector measures 1.20 m in diameter, corresponding to an aperture area of roughly 2 m^2 and a depth of about 0.3 m. Depending on how carefully the parabola was molded and how accurately the mirror fragments were installed, the diameter of the focal spot amounts to about 10 cm.

This model took the second prize in Chile's 1987 solar cooker competition /65/. One of its main advantages is that it can be built at little expense (but with a lot of work). Unfortunately, in addition to having the standard disadvantages of conventional reflector cookers, it is also somewhat awkward to operate: to stir the food, for example, the cook has to bend over the reflector to reach the swaying pot at the device's focal spot.

<u>Figure 38:</u> **Valparaiso Reflector Cooker**

Designation: VIAX Solar Cooker SK 10
Type: Reflector cooker
Invented/ developed by: D. Seifert, Siedlungsstr. 12 D-8265 Neuötting, FRG

The VIAX Solar Cooker is a customer-assembled metal reflector cooker with an especially sophisticated type of tracking mechanism (see figs. 39 and 40).

The reflector is vertically (2) and horizontally (3) adjustable. As it is turned by hand around its vertical axis in the course of a day (azimuthal adjustment), the proper horizontal adjustment with respect to the sun's elevation is achieved automatically by way of a fixed guide element (5), the constant length of which ensures that the reflector is aimed skyward when pointing to the south, gradually tilting downward when moved toward the east or west (depending on the time of day). Altogether, this corresponds to revolution around a virtual axis (hence the name) - the polar axis. A seasonal adjusting mechanism (7 and 8) enables accommodation of seasonal variation in solar altitude. Tracking is necessary only every 25 min.

Appropriately, the VIAX mechanism is installed behind the reflector. The journal "Sonnenenergie" 10, 3, 1985 includes a detailed description. For checking and testing the food, the SK 10 parabolic mirror can easily be turned away without moving the cooling vessel in the pot holder. Consequently, there is no danger of dazzling, blinding, or burning. The rest and night position of the reflector is upside down for dust protection. Several VIAX cookers have been tested so far.

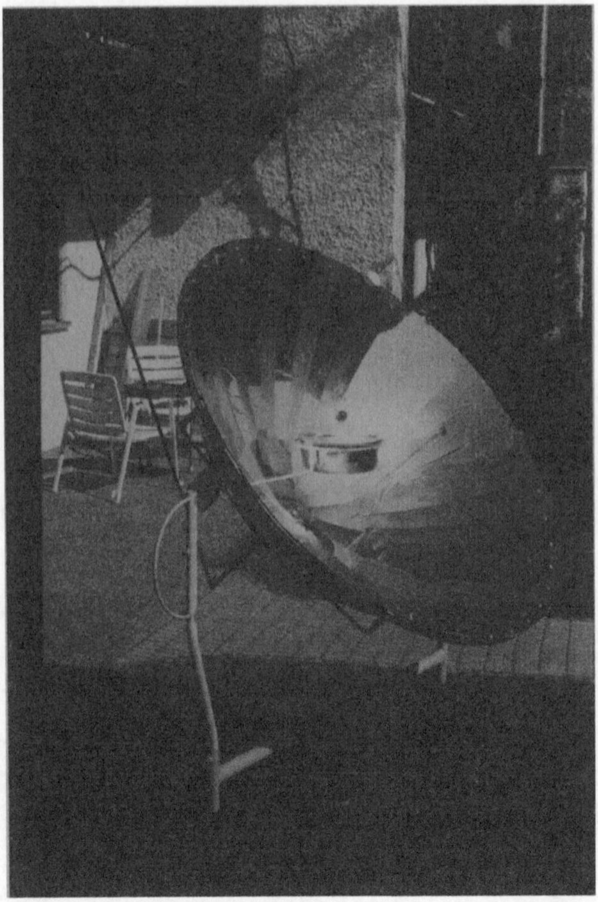

Figure 39: VIAX Solar Cooker SK 10

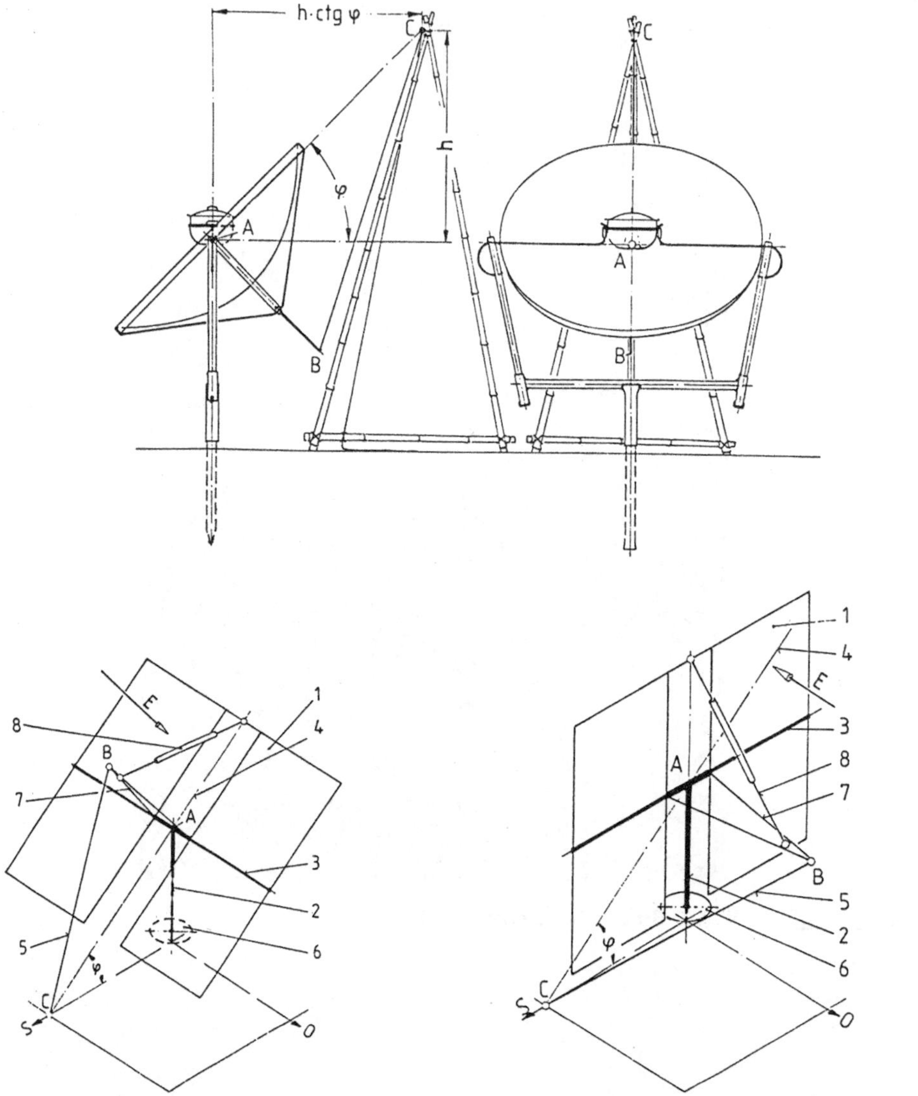

receiver plane in equinoctial noon position

receiver plane in equinoctial sunrise position

Figure 40: VIAX Solar Cooker - tracking geometry

<u>Heat-accumulating solar cookers</u>

Designation:	Heat-accumulating Steam Cooker
Type:	Linear-parabolic reflector cooker with heat storage in steam and water
Developed by:	David R. Mills, School of Physics, University of Sidney, Australia Mao Yin Qiu, Gansu Natural Energy Research Institute, Lanzhow, Gansu Province, China

Compared to most other solar cookers, the solar cooking stove, or solar steam cooker, is a relatively sophisticated device that unites some of the characteristics of reflector cookers, steam cookers and heat-accumulating solar cookers.

The principle of operation: Water is heated and subsequently evaporated in an evacuated glass absorber tube situated in line with the focal line of a linear parabolic reflector. The water is transported to a barrel-shaped, water + steam-filled heat store by natural (thermosiphon) circulation; as more water evaporates, the pressure increases steadily. The top of the cylindrical heat store serves as a hot plate that imparts the heat of the steam to the cooking pot/food. The process involved is similar to that of a heat pipe: steam condenses on the relatively cool surface of the pressure vessel and gives off latent heat (of condensation). That causes the pressure to decrease, consequently allowing more water to evaporate, and the process continues.

According to Mills and Mao Yin Qiu, the maximum achievable temperature is about 170°C. The system is designed for a corresponding watervapour pressure. The device can store about 2 kWh per day of good sunshine, or enough to prepare about 50 sausages or 50 fried potatoes. At 4 kW, the starting power is relatively high. The linear parabolic reflector has to be adjusted to the seasonal solar altitude 10-20 times a year. Daytime solar tracking is unnecessary.

The reflector area is 2 x 0.9 m^2, and the storage volume is 23 l.

A solar cooking stove in combination with a linear reflector constitutes a totally new combination of ideas and action principles in solar cooking. However, it is comparatively expensive. One prototype has been built to date. A detailed description along with test data is contained in /110/.

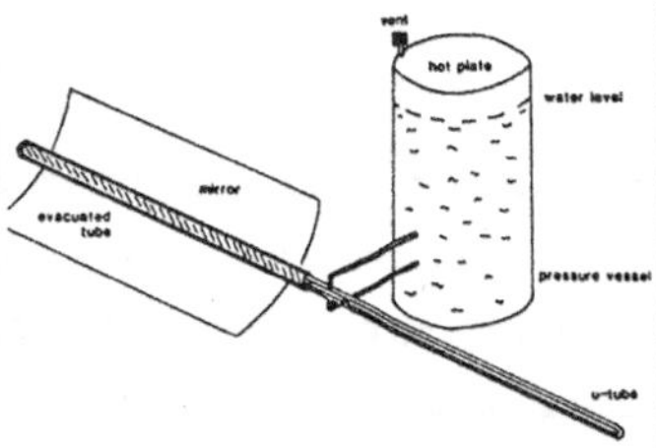

Internal solar cooking stove construction is shown. The U-tubes are fitted with conducting fins (not shown) to enhance heat transfer from the evacuated glass absorber tubes.

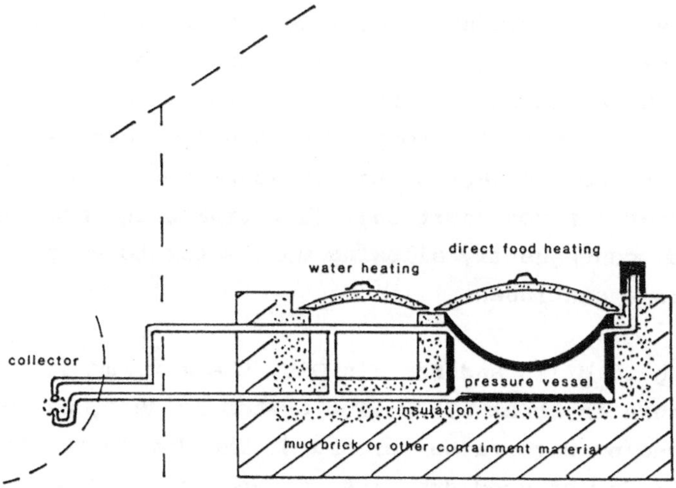

SOLAR COOKING STOVE

A sketch of a dual hotplate design incorporating two insulated heating cavities. The left cavity is unpressurised and uses connecting pipes to and from the collector and storage vessel for heating of water. The right cavity is above a pressure vessel and is used for high temperature direct heating of food.

Figure 41: **Heat-accumulating Steam Cooker**

Designation: Hybrid fixed-focus cooker

Type:

Invented/ developed by: W. Scheffler, c/o Oehler, Morgartenring 18, CH-4054 Basel, Switzerland
Ministry of Energy and Resources Nairobi, Kenia

Big combined solar and fuelwood cooking device for institutional use in schools, etc.

The cooker consists of 2 parabolic reflectors and a central unit with 2 glass windows of 0.6 m diameter. Reflected sunlight from the parabolics passes the windows and is reflected by two cured aluminum secondary reflectors to strike sides and bottom of the pot.

In case the sun is not shining, the secondary reflectors can be removed and replaced by a fuelsaving woodstove under the pot for conventional cooking.

Technical data:
2 parabolic reflectors of 3.8 m^2 each with 21 flat segments in an ellipitc frame of 2 m x 2.7 m.
Total aperture 5.5 m^2, focal length 2.0 m.
Focus diameter 0.6 m.
Pot diameter 0.5 m, volume 54 l.
Effective heating power 1.6...2.2 kW (at clear sunshine 900 m above sea level, depending on pot temperature), heat losses of the pot at 100°C approx 0.5 kW.
Estimated thermal efficiency 35...45 %.

Prototype constructed in 1987; a second item with 2 pots of 100 l each is under construction.

Designation: ISE Solar Cooker with Integral Oil Storage

Type: Heat-accumulating solar cooker with flat collector and fluid heat store

Developed by: Fraunhofer-Institut für Solare Energiesysteme (ISE), Oltmannstr. 22, D-7800 Freiburg, FRG

Manufactured by: Cambra GmbH, D-7260 Calw, FRG

The ISE Solar Cooker with Integral Oil Storage is a heat-accumulating type with numerous virtues. First of all, it will cook food anytime day or night, has medium-high temperatures and reasonable cooking times. Additionally, the cook needn't stand out in the hot sun; indeed, the "range" part can even be installed indoors. The device exploits both direct and diffuse radiation, requires no solar positioning, allows access to the pot during cooking, and is easy and simple to operate. On the other hand, those numerous advantages are obtained at the cost of technical complexity and a very high price: Units from short-series production cost between US$ 1250 and US$ 1500. Consequently, though an early version with a 1.9-m^2 aperture held 5 kg food, and the capacity of a later model is 10 kg, the cooker is still disproportionately expensive for its size.

A closed-loop heat cycle in which oil transfers thermal energy from a highly efficient flat plate collector to an elevated heat-storage vessel (oil storage), serves as the device's functional basis. The thermal oil serves both as the energy vehicle and the heat store. Circulation between the collector and the oil storage tank requires no pump, since it is based on natural circulation, i.e. the thermosiphon effect. The cooking pot is placed in a cavity in the top of the oil storage tank. Heat losses are minimized by an insulating cover that can be raised during cooking, thus allowing access to the food for stirring etc. without wasting too much heat.

For an ambient temperature of 25°C, the cooker can achieve a maximum temperature of 160°C. The 50-l oil store will retain enough heat to cook two consecutive batches of food (3.5 to 4 kg each).

About 20 prototypes have been built to date and are being tested in various parts of the world. Ongoing developments continue to aim for larger units (for use at the institutional level) with more efficient collectors, local producibility and, above all else: lower prices.

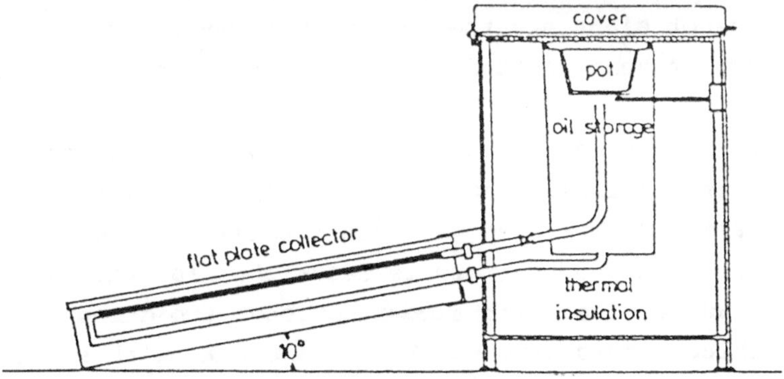

Figure 42: ISE Solar Cooker with Integral Oil Storage

Designation:	Solar Hot Plate Cooker
Type:	Heat-accumulating solar cooker with radiation-intensifying reflectors in split construction
Developed by:	Bomin Solar, Industriestr. 8, D-7850 Lörrach, FRG

The Solar Hot Plate Cooker is the only solar cooker ever to be strictly designed as a split device. It consists of two units: one for collecting the radiation and converting the energy into heat, and the other for storing the heat and doing the actual cooking. Collecting the radiation and cooking are therefore temporally and spatially separate functions, the advantage being that meals can be cooked indoors and whenever so desired. The collector essentially consists of a rather large box type solar cooker into which the solar radiation impinging on a reflecting funnel made of 8 flat mirrors is directed at 3.5-fold intensity; passing through 3 sheets of fluoropolymer foil, the incident energy heats up a set of selectively coated iron slabs. The whole device is mounted in a tracking mechanism that allows hourly repositioning with the aid of a crank.

When the slabs are hot enough (200...300°C), a transport box is used to remove them from the collector box through an opening in the side and transfer them to the indoor heat-storing stove. The latter is a very good heat retention. For cooking, one merely folds it open and places the pot(s) on the hot slabs of iron. The device has a very high starting power. The contact between the hot slabs and the cooking pot can be varied to yield a sort of heat control for fast or slow cooking.

In addition to its main advantage of making the cooking process temporally and spatially independent of energy collection, the Solar Hot Plate Cooker also has some interesting design details: the funnel of mirrors reflects the radiation at Brewster's angle, i.e. with almost no loss; two-dimensional solar tracking occurs once per hour via crank, with a sighting instrument provided for easy checking of the adjustment; the heatproof plastic cover costs less and yields higher efficiency than two panes of glass.

The drawbacks of a Solar Hot Plate Cooker are its elaborate design and somewhat complicated handling and operation. Controlling the thermal output of the cooker requires experience and a delicate touch. If food boils over or a pot is spilled, subsequent cleaning can be problematic. For this reason, an additional joint around the pot was provided to catch the food spilled over.

At a large-series price of US$ 400 (individual local production: US$ 800), the Solar Hot Plate Cooker is less expensive than other heat-accumulating cookers, but still too expensive to gain widespread acceptance in developing countries.

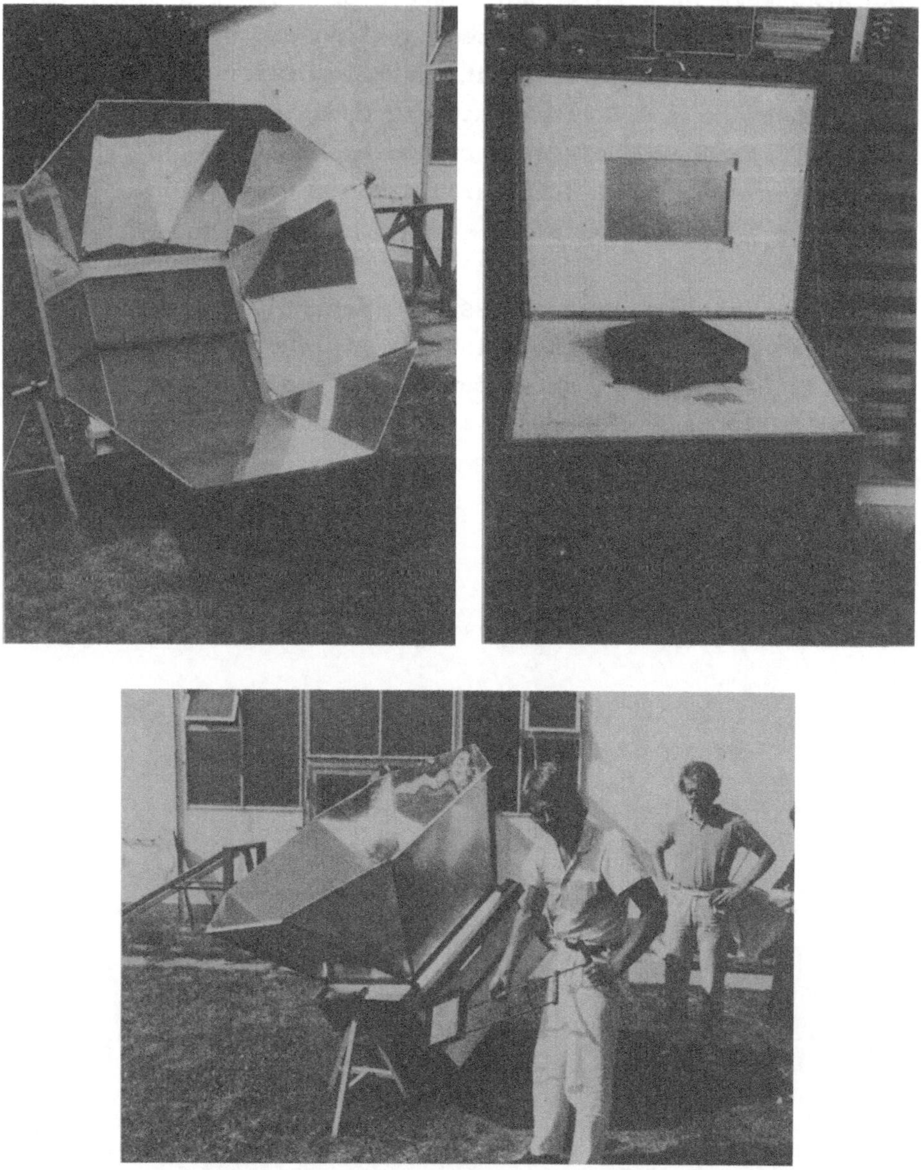

<u>Figure 43:</u> Solar Hot Plate Cooker. Clockwise from top left: front view of collector; cooking/heat-storage unit (open); handling a hot slab

Designation:	Heat-pipe Storage Cooker
Type:	Heat-accumulating solar cooker with evacuated tube collectors
Invented by:	Stoy/Pöhlmann
Developed/ manufactured by:	AEG, Geschäftsbereich Hausgeräte Muggenhofer Str. 135, D-8500 Nürnberg 80, FRG

This solar cooker consists of a heat-storage block under a hot plate and lid, with an inclined bank of 6 evacuated collector pipes in a reflective tray with a transparent protective cover, all arranged as shown in figure 45.

The evacuated tube collectors function according to the heat-pipe principle (modified Philips collector tube); the heat collects at the head of the tube, from where it is directly transferred to the heat store without need of a special circulating loop.

The heat store is made of magnesite (the prototypes contained 50 kg aluminum). The combination hot-plate + cooking cavity arranged above the heat store takes its heat through a second short high-temperature heat pipe. Heat control is provided by a valve in the heat pipe.

The device has essentially the same advantages as those inherent to any heat-accumulating type of solar cooker: cooking is possible anytime day or night, and the cook needn't stand out in the sun. Additionally, it is smaller and lighter than the other two types of heat-accumulating solar cookers described above and can be pulled along on its two wheels (or skids if necessary). Though still in the development stage, its external design is already practically perfect.

The hot plate requires the use of flat-bottomed pots (almost unknown in developing countries); since the pot is only heated from the bottom, the hot plate must develop a higher temperature than would be necessary in a cooking cavity or box-type solar cooker, and the size of the hot plate leads to substantial heat losses around the pot. Under such conditions, the collecting/storage capacity is relatively modest: the heat store can deliver about 1 kWh for the useful temperature range of 140...220°C.

To make use of at least part of the waste heat, the insulated cover has been replaced by a dome that covers cooking plate, cooking mould, and cooking pot.

According to the manufacturer, the device could be improved by installing evacuated collector pipes that function at higher temperatures and by intensifying the transfer of heat from the heat pipe to the heat store.

The two main problems are a) inefficient transfer of heat to the cooking vessel and b) sophisticatedness and price (approx. US$ 1750, or half that much if mass-produced), the latter factor having the effect of precluding use of the Heat-pipe Storage Cooker in developing countries.

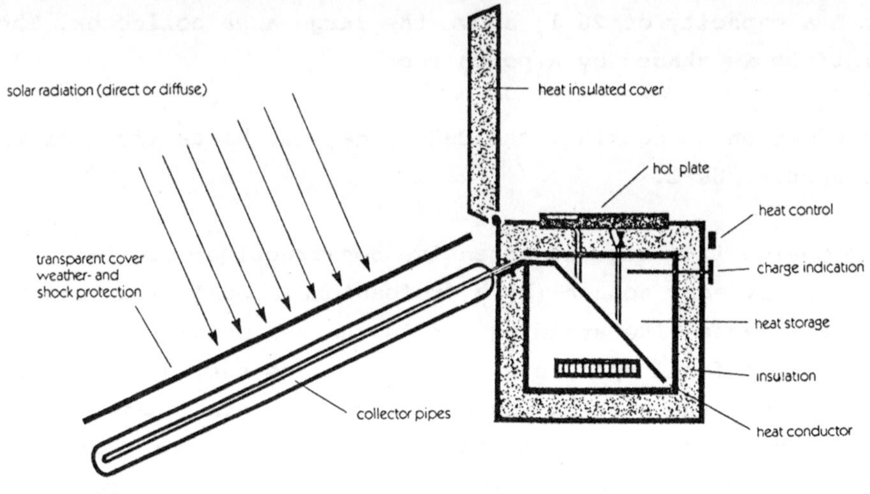

<u>Figure 44:</u> Heat-pipe Storage Cooker with Evacuated Tube Collectors

<u>Convective solar cookers</u>

Designation:	Convective Solar Cooker (CSC)
Type:	Large-scale solar range with air convection
Invented/ developed by:	Synopsis, Institut de Recherche Alternative, Domaine de Belbezet, Route d'Olmet, F-34700 Lodève, France

The Synopsis Convective Solar Cooker has an aperture area of nearly 10 m^2. Its design capacity is 182 l, corresponding to a warm meal for roughly 200 people.

Heat transfer is effected mainly by air convection, with the air heating up in a flat plate collector, ascending to the elevated cooking pots and heating them from all sides. Insulation onto the flat plate collector is intensified by a reflecting mirror.

The cook stands on a platform to tend 7 (!) large vessels (each with a capacity of 26 l) above the large-area collector. The platform is shaded by a porch roof.

In addition to cooking, the CSC range also heats 150 l of water to approx. 80°C.

The device is a big and expensive construction, but is relatively easy to use (by a professional cook). The "classic" disadvantages like standing out in the sun, handling pots in the unit's focal spot, etc. have been done away with. On the other hand, the size of the cooker and its stationary installation preclude optimized solar tracking for a particular time of day (morning or afternoon). Though air is a poor heat conductor, the device is reported to be relatively efficient (η = 43 %).

The only large-scale CSC ever commissioned is in operation at a school in the Sudan, where it is used to cook for 250 people. It is intended to install more, improved CSCs in other places.

Since the device was designed for use by professional cooks at the institutional level, the question of acceptance by the users reads differently than in the case of family-size cookers. An institution can be expected to base its investment decisions on more long-range planning than an individual family would, and the cooking staff can be expected to have been both properly familiarized with the cooker and trained in its proper use. Under comparable circumstances, some types of solar cookers must be regarded as acceptable, while they would have no chance at all of being accepted by individual families.

Only one CSC has been built to date, and it is still undergoing field and acceptance testing. There were also CSCs of smaller size developed by Synopsis. The series production price 20 to 500 units per month of the "family model" has theoretically been calculated to US$ 152 to 582, for the 50 persons module US$ 223 to 698.

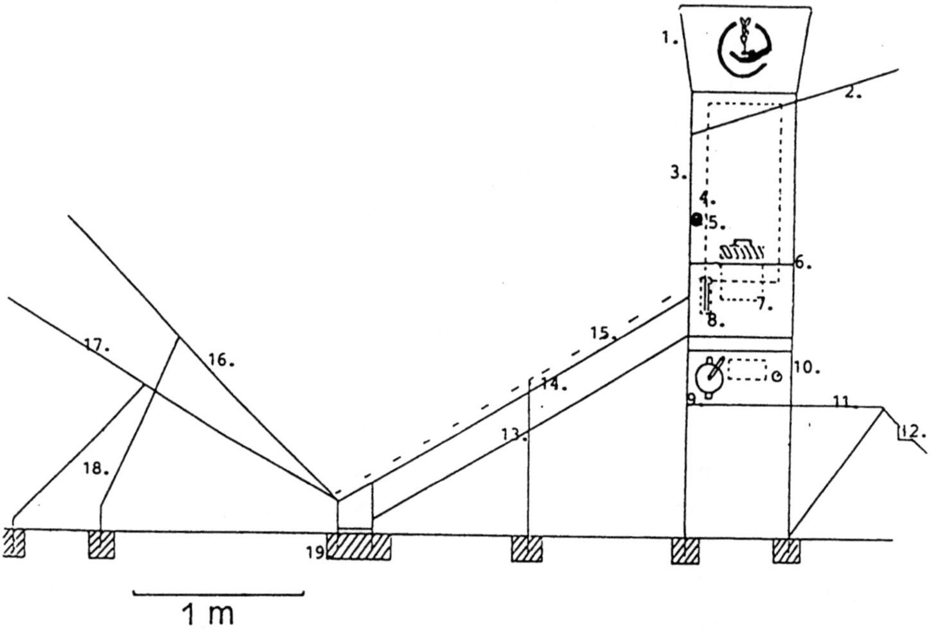

Schematic diagram of the CSC in Kadugli

(1) Cold water tank, (2) roof of the kitchen range, (3) vertical reflector, (4) hot water tank, (5) hot and cold water taps, (6) work table with inset pots (7), (8) heat exchanger for the hot water storage, (9) hand pump for the cold water storage, (10) manometer and secondary circuit accessories, (11) platform, (12) steps, (13) convective solar collector, (14) support for covers in the closed position (15), (16) in the March to October position, (17) in the November to February position, (18) cover supports, (19) foundations

Figure 45: Convective Solar Cooker (CSC)

Designation: Steam Immersion Heater Solar Cooker

Type: Closed-loop steam s.c. with high efficiency collector

Developed by: V. Heinzel, J. Holzinger
Institut für Reaktorentwicklung,
Kerforschungszentrum Karlsruhe,
Postfach 3640, D-7500 Karlsruhe, FRG

Solar collector tube, steam separator, riser and downcomer tubes, air well, siphon tubes, and immersion heater together make up a closed circuit, partly filled with water. Solar radiation striking the collector tubes generates steam which rises up to the immersion heater inside a cooking pot. Heat passes from the steam to the food, thus condensing the steam to water. A siphon tube evacuates the condensate via the downcomer back to the collector tube. An additional siphon breaker tube is installed to avoid water-air pulsations in the system.

As the immersion heater is attached to the circuit with flexible hose pipes, it can be used for indoor cooking in any type of cooking pot. Providing that the pot is located 50 cm higher than the collector, the circuit runs spontaneously without any auxiliary energy.

Allowing indoor cooking with different types and sizes of cooking vessels, the system is extremely flexible and comfortable. Because of the closed circuit, higher pressure and higher temperatures are possible than with atmospheric steam cookers, leading to better heat transfer and shorter cooking times.

This s.c. is a high-cost solution requiring high efficiency collectors (e.g. evacuated tubular collectors) and an accurately assembled immersion heater and siphon tube system.

Till 1988, one prototype was realized to demonstrate the physics of the self-acting siphonic heat transfer system. (See also ref. /195/).

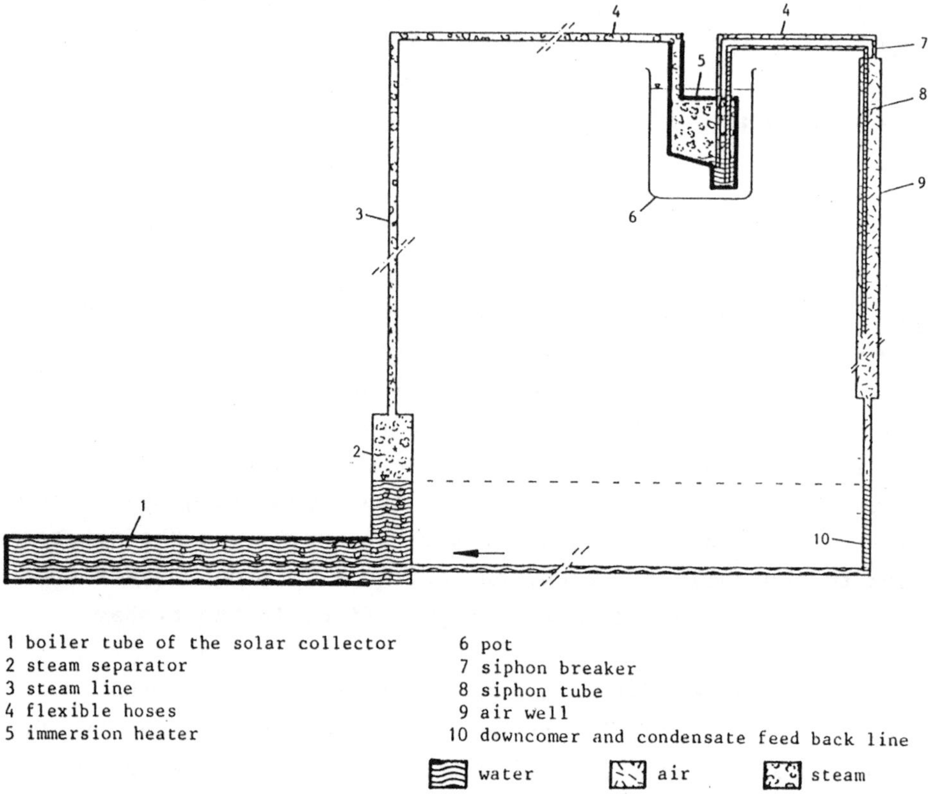

Figure 46: **Steam Immersion Heater Solar Cooker (from /194/)**

Comparative survey

Designation	ULOG Tropical 85	ULOG European	Indian Box Type Solar Cooker	Suriamuki Box Type Solar Cooker	ATRC Solar Cooker
Designed/built by	U.Oehler, Basle	U.Oehler, Basle	various manufact.	Centre of Energy	ATRC Bardoli Ind.
Marketed by	local craftsmen	carp.+ homeworkers	local dealers	Studies IIT Delhi	Yantra Vidyalaya
Qty. produced	450 (incl.pr.mod.)	1.600	~ 100.000	just beginning	1.000 since 1977
Type of construction	solar cooking box with reflector	solar cooking box with reflector	solar cooking box with reflecting lid	solar cooking box with lid; small, lightweight	solar cooking box with reflector
Dimens.: LxWxH (cm)	67 x 67 x 50	-	58.5 x 58.5 x 20	40 x 40 x 18	76 x 46 x 20
Aperture (cm x cm)	47 x 47	43 x 43	approx. 50 x 50	30 x 30	0.21 m^2
Concentr.factor	approx. 1.5	none	approx. 1.5	1.5	approx. 1.5
Intensify.mirror	1 (lid)	1 (lid)	1 (lid)	1 (lid)	1 (lid)
Weight (kg)	approx. 9 kg	no info.	10-12 kg	7 kg	16 kg
Portability	bulky	somewhat bulky	"suitcase", rollers	very good	like heavy suitcase
Heating power (1 l H_2O, 20-100°C)	no reliable info.	-	2 liter:90-100min. at 7 W/m^2	no info.	no info.
Efficiency (partial effic.)	approx. 45 %	-	approx. 50 %	30 % at 650-900 W/m^2	no info.
Max. temperature	180°C (35°C amb.)	no info.	140-150°C (20°C amb.)	100°C (?) (17°C amb.)	150°C (38°C amb.)
Qty. of food (kg)	3 kg	similar	2.5 kg	2.5 kg	3 kg
No. of vessels	4	to ULOG	4	4	4
Tea/baked goods	yes/yes	Tropical 85	yes/limited	yes/yes	yes/yes
Roasting/grilling	no/no		no/no	no/no	no/no
Keeping warm	approx. 2-3 h	↓	4-5 h	approx. 4 h	approx. 2 h
Must cook stand in sun?	no		no	no	no
Stirring possible?	no		no	no	no
Stirring neces.?	no		no	no	no
Solar tracking?	no		no	no	no
Flare?	no		no	no	no
Blinding?	no		no	no	no
Burns?	no		no	no	no
Special disadvantages	long cooking time, rain damages box		long cooking time, low temperatures	long cooking time, low temperatures	heavier than other solar cooking boxes
Special advantages	no attendance, no burnt food, no spilling over		no attendance, no burnt food, no spilling over	cheap, small, lightweight	rectangular shape good f. stability and energy capture
General handling	good, easy, simple		very good, easy, lightweight, simple	good and easy thanks to smallness	easy, especially on revolving stand (option)
Manufacture	easy		handicraft	handicraft	handicraft
Requisite handicraft skills	metalcutting, woodworking, sawing, nailing		metalworking, woodworking	metalworking	glazing and coating
Requisite tools and fixtures	hammer, saw, tin snips, file, pliers		normal workshop tools	normal workshop tools	normal workshop tools
Materials not locally available	possibly glass		inner cover made of safety glass	none	possibly aluminum sheet
Weak points	glass cover		glass, hinges	cover + mirror	glass
Remedy by	replace by hand		replace by hand	replace by hand	replace
Service life	5-6 years		10-15 years?	15 years (?)	5 years
Cost of manuf. US$	$ 35-40, 16 man-h.		$ 50	$ 23	$ 50
Purchase price US$	no info.		$ 25	$ 25	$ 60
Cost of maint. US$/a	$ 5		$ 5	$ 5	$ 5
PEL (USc/kWh)	4.0 c		8.8 c	27 c	no info.
CC, CCL (USc/kg)	3.3 c (CC)		4.9 c (CC)	2.5 c (CC)	no info.
Target group (acc. to manuf.)	urban and rural middle class		urb.mid.+upper cl. and poor rural population	urban + rural, all classes	urban middle class, rural lower class

Designation	Solar Cooking Box Sudan SEP	ATDO Reflector Box Type Solar Cooker	22° Solar Cooker Zimbabwe	Dhauladhar Solar Cooker	RIIC Solar Oven
Designed/built by	VTC Khartoum (ERC)	ATDO, Pakistan	I.de Klerk, Harare	I.G.Dhauladhar	RIIC, Kanye,
Marketed by	no info.	ATDO	no marketing	Project	Botswana
Qty. produced	100 since 1986	approx. 20	21 since 1984	approx. 150 since 1985	30 since 1981
Type of construction	solar cooking box with reflector	solar cooking box with reflector	solar cooking box with reflector	adobe solar cooking box	solar cooking box with 4 reflectors
Dimens.: LxWxH (cm)	75 x 55 x 35	76 x 59 x 23	-	170x75x75 (outside)	50 x 45 x 40
Aperture (cm x cm)	0.35 m^2	0.45 m^2	-	60 x 90	0.1 m^2 (no refl.)
Concentr.factor	approx. 1.5	approx. 1.5	approx. 1.3	none	approx. 3
Intensify.mirror	1 (lid)	1 (lid)	foil refl.in lid	with or without	4 reflectors
Weight (kg)	8 kg	15 kg	4 kg	no info.	approx. 9 kg
Portability	light	satisfactory	light	stationary adobe	portable
Heating power (1 l H_2O, 20-100°C)	no reliable info.	2.5 liter: 26°C to 100°C in 60 min., no radiation data	no reliable info.	1 liter H_2O in 30 min.	1 liter: 30-100°C in 240 min. 700 W/m^2 (30°C amb.)
Efficiency (partial effic.)	no info.	approx. 55 %	no reliable info.	approx. 40 %	29 % without refl. 10 % with reflect.
Max. temperature	150°C (37°C amb.)	170°C (30°C amb.)	no info.	160°C	175°C
Qty. of food (kg)	5 kg	4-5 kg	4 kg	6 kg or more	(6?) 3 kg
No. of vessels	4	4	2	6	1-2
Tea/baked goods	yes/yes	yes/yes	yes/yes	yes/yes	yes/yes
Roasting/grilling	yes/no	yes/no	yes(?)/no	yes/yes	no/no
Keeping warm	approx. 5 h	no (?)	yes	3-4 h	no
Must cook stand in sun?	no	no	no	no	no
Stirring possible?	no	no	no	no	no
Stirring neces.?	no	no	no	no	no
Solar tracking?	no	no	rarely	no	every 1/2 h
Flare?	no	no	no	no	no
Blinding?	no	no	no	no	no
Burns?	no	no	no	no	no
Special disadvantages	long cooking time	no info.	no info.	damaged by rain	PU insulation not durable enough
Special advantages	no attendance, no burnt food, no boiling over	"drawer" type opens at the side	no info.	large enough for big families	no info.
General handling	good, simple	practical	handle glass and reflector with care	good	good
Manufacture	handicraft	by craftsmen	easy for craftsmen	easy for craftsmen	easy
Requisite handicraft skills	carpentry	metalworking	no special skills	masonry, glazing precision work	sheet metalworking
Requisite tools and fixtures	saw, hammer, pliers, screwdriver, tin snips	normal workshop tools	no special tools	none	no special tools
Materials not locally available	none	aluminum sheet?	none	none	none
Weak points	glass	glass breakage	no info.	adobe in monsoon season	glass
Remedy by	replace by hand	replace by hand	no info.	repair by hand (housewife)	replace by hand
Service life	5 years	no info.	(20?) 5 years	10 years? (contin. repair)	2 years
Cost of manuf. US$	approx. $ 37	$ 24	$ 30	$ 22 for material	$ 45 (1981)
Purchase price US$	approx. $ 50	$ 30	$ 50	without work	$15
Cost of maint. US$/a	$ 5	$ 5	$ 5	$ 2	$ 5
PEL (USc/kWh)	-	4.8 c	-	2.1 c	47 c
CC, CCL (USc/kg)	-	1.6 c (CC)	-	0.9 c (CC)	3.7 c (CC)
Target group (acc. to manuf.)	urban and rural low-to-mid class	urban middle class	urban middle class	subsistence farmers at higher elevations	village families

Designation	SERVE Box Type Solar Oven	EEB Solar Cooker	Falco S/C	Fixed-Focus Hybrid Solar Cookers	Sun Basket
Designed/built by Marketed by	SERVE, Peshawar, Pakistan	Energy Res. Inst. Henan Acad.of Sci.	BBK J.D.Kaller&P. Kleingsenget, FRG	W.Scheffler, Nairobi	ICRISAT, v.Oppen Walker, Hyderabad
Qty. produced	2000 since 1985	100.000 since 1975	no info.	3 since 1986	20 since 1985
Type of construction	s.c.box with inclined cover and reflector	eccentric axis refl. cooker	reflector cooker with medium focal length	hybrid parabolic reflector and wood fire cooker	reflector cooker with short focal length
Dimens.: LxWxH (cm)	96.5 x 52 x 40		115 cm dia.		130 cm dia.
Aperture (cm x cm)	0.39 m^2 (without refl.)	1 m^2	1.04 m^2	7.6 m^2; 22.4 m^2	1.3 m^2
Concentr.factor	approx. 1.5	approx. 80	approx. 10	6.7; 7	approx. 10
Intensify.mirror	1 (lid)		alum.rot.parab.	2; 4 (refl.foil)	fibergl.rot.parab.
Weight (kg)	approx. 28 kg	wood shell,mirrors	less than 20 kg		8 kg (no stand)
Portability	easy for 2 people	appr. 15 kg	bulky, very light	no	bulky but light
Heating power (1 l H_2O, 20-100°C)	2 liter: 19-100°C in 150 min. at 658 W/m^2 (25°C amb.)	collapsable,wheels 1 liter: 15-100°C in 13 min. at 720 W/m^2	2 liter: to 100°C in 30-45 min., 985 W/m^2	54 l; 200 l 60-100°C 90 min.	1 liter: 25-100°C in 9 min.
Efficiency (partial effic.)	29 %	63 %	25-35 %	28 %; 31 %	approx. 50 %
Max. temperature	175°C (45°C amb.)	appr. 500°C	no reliable info.		400°C (30°C amb.)
Qty. of food (kg)	4 kg		no info. (5 kg?)	54 kg; 200 kg	3 kg
No. of vessels	4	1	2	1; 2	1
Tea/baked goods	-/yes	yes/possible	yes/yes	yes/no	yes/yes
Roasting/grilling	no/no	yes/yes	yes/yes	no/no	yes/no
Keeping warm	approx. 2.5 h	no	no	yes	no
Must cook stand in sun?	no	yes	yes	no	yes
Stirring possible?	no	yes, high level	yes, difficult	yes	yes, awkward
Stirring neces.?	no	yes	yes	yes	yes
Solar tracking?	no	yes	yes, every 10 min.	yes	yes
Flare?	no	no?	yes	yes	yes
Blinding?	no	no?	yes	yes	yes
Burns?	no	yes	yes	yes	yes
Special disadvantages	glass breakage, foil damage	high pot level	stand, diffusely reflected radiation		pot hard to reach in focal spot
Special advantages	low-temperature cooking without attendance	fast cooking, high efficiency, transportability	fast cooking, easy adjustment	hybrid: wood fire possible	fast cooking
General handling	good (after proper familiarization)	easy	easy		not so simple
Manufacture	by craftsmen	craftsmen, industry	industrial	by craftsmen	handicraft
Requisite handicraft skills	carpentry	joinery	pressing (mirror), finishing, welding	metalworking, bricklaying	build mold, shape fiberglass, welding
Requisite tools and fixtures	circular saw tin snips	joiner shop tools	very large press, coating machine, welding unit	machine-shop tools	shop tools, welding unit
Materials not locally available	silicone, safety glass	none	reflector, special pots	reflecting foil	possibly fiberglass
Weak points	glass,paint,foil	reflector	reflector finish	foil reflectors	mechanism,coating
Remedy by	-		no info.	new foil	by hand, apply new reflective coating
Service life	no info., 5?	no info.	no info.(10years?)		10 years
Cost of manuf. US$	$ 63	no info.	no info.	$ 2,250; $ 3,350	$ 80
Purchase price US$	$ 18	no info.	$ 250 (in lots of 100 items)		$ 40
Cost of maint. US$/a	$ 5	no info.	$ 10		$ 15
PEL (USc/kWh)	15.3 c	no info.	18 c		5.2 c
CC, CCL (USc/kg)	3.5 c (CC)	no info.	9.3 c (CC)	7.7 c; 3.1 c	3.4 c (CC)
Target group (acc. to manuf.)	refugees in camps	rural people, farmers, etc.	no info.	school children	urban low-to-middle class

Designation	Sungril	Tube Solar Oven	VIAX Solar Cooker SK 10	Cylindrical Parabolic Solar Oven	Cylindrical Parabolic Solar Cooker
Designed/built by	Interel S.A.	K.Wippermann,	D.Seifert, Neuöttg.	Tharwat Sabry	E. Floegel
Marketed by	Basel, Switzerland	Univ. of Karlsruhe	no marketing	Cairo	Hannover, FRG
Qty. produced	no info.	1 prototype	> 3 since 1986	10 since 1954	1 prototype
Type of construction	collapsable reflector cooker	linear parabolic reflector s.c. with cooking/baking tube	reflector cooker	parabolic reflector	cylindrical-parab.
Dimens.: LxWxH (cm)	100 cm dia.	122 x 74 x 44	120 x 160 x 160	250 x 60 x 100	300 x 130 x 115
Aperture (cm x cm)	0.78 m^2	0.6 m^2	1.1 m^2	1.5 m^2	4 m^2
Concentr.factor	45	2.4	approx. 10	17	10
Intensify.mirror	alum.rot.parab.	lin.parabol.mirror	metal.parab.refl.	alum.paraboloid	polystyrene/alum.
Weight (kg)	6 kg	18 kg	< 30 kg	15 kg	no info.
Portability	easy, handy	by 1 person	disassemblable	light, disassembl.	disassemblable
Heating power (1 l H_2O, 20-100°C)	550 W 1 liter: 20-100°C in 10 min.	4 liter: 200-100°C in 83 min, 833 W/m^2	3 liter: 20-100°C in 45 min. at 700 W/m^2 (25°C amb.)	4 liter: 17-100°C in 30 min. no rad. info. (35°C amb.)	10 liter: 25-100°C in 45 min. 800 W/m^2 (21°C amb.)
Efficiency (partial effic.)	approx. 70 % (?)	46 % (rel. to global radiation)	34 %	approx. 60 %	36 %
Max. temperature	450°C	165°C at 20°C amb. temp. and 880 W/m^2	no info.	450°C (35°C amb.)	150°C (20°C amb.)
Qty. of food (kg)		4 kg	no info. (5 kg?)	4 kg	30 kg
No. of vessels	1	1 or 2 spec.vessels	1	1	4
Tea/baked goods	yes/no	yes/yes	yes/possible	yes/yes	yes/yes
Roasting/grilling	yes/yes	yes/no	possible/no	yes/yes	yes/no
Keeping warm	no	no	no	3 h	no
Must cook stand in sun?	yes	partly	yes	yes	no info.
Stirring possible?	yes	partly	yes	yes	yes
Stirring neces.?	yes	partly	yes	yes	yes
Solar tracking?	yes	every 40 min.	yes, every 25 min.	yes, every 40 min.	no
Flare?	yes	little	yes	yes	yes
Blinding?	yes	little	yes	yes	yes
Burns?	yes	little	yes	yes	yes
Special disadvantages	not very stable, sensible to wind	glass delic., spec. pots required	operation requires prior instructions	special pot, no trad cooking vessels	requires adept operation/handling
Special advantages	all aluminum, easy transport	simple mirror, high efficiency, easy tracking	simple tracking, easy maintenance	simple cooking and handling	infinitely adjustable thermal output
General handling	easy	easy mounting and loading	easy, assembly, requ. experience	easy	no info.
Manufacture	by industry	by craftsmen	handicraft	simple	by craftsmen
Requisite handicraft skills	diff. kinds of aluminum metal-working	simple joining and metal working, glass-blowing	metalworking	welding metalworking	no info.
Requisite tools and fixtures	industrial equipment	joining, sheet metal working, and glass blowing tools	drill, welding unit, bending tools	normal village workshop tools	metal-cutting saw, drill
Materials not locally available	polished aluminum reflector sheets	reflector, glass tube	reflective foil, pot with glass cover, polystyrene vessel	reflecting sheets?	aluminum sheet for reflector
Weak points	aluminum reflector	tube, reflector	cable guide, mirror	reflector	no info.
Remedy by		replace tube/mirror	by hand	trained personnel	no info.
Service life	no info.	no info.	(20yrs?) 10 yrs.	(15?) 10 yrs.	no info.
Cost of manuf. US$	no info.	approx. $ 200	(65?) at least $130 in series	approx. $ 160	no info.
Purchase price US$	no info.	no info.		no info.	no info.
Cost of maint. US$/a	no info.	no info.	$ 20	$ 15	$ 20
PEL (USc/kWh)	no info.		20 c	5.7 c	-
CC, CCL (USc/kg)	no info.	9.3 c	4.8 c (CC)	7.4 c (CC)	-
Target group (acc. to manuf.)	no info.	no info.	no info.	campers/vacationers	no info.

Designation	ISE Solar Cooker with Oil Storage	Solar Hot Plate Cooker	Heat-pipe Storage Cooker	Convective Solar Cooker - CSC
Designed/built by	Fraunh.ISE,Freibg.	Bomin Solar	Poehlmann/Stoy	Synopsis, Lodeve F
Marketed by	Cambra, Calw, FRG	B.S.Lörrach, FRG	AEG	AICF
Qty. produced	20	approx. 10	approx. 4	prototypes
Type of construction	heat-accu. solar cooker with flat collector and fluid heat store	slightly concentrating split-type, heat-accu. solar cooking box	heat-accumulating solar cooker with evac. collector pipes	large-scale cooker with air convection
Dimens.: LxWxH (cm)	230 x 150 x 110	200 x 160 x 160	210 x 55 x 50	500 x 330 x 280
Aperture (cm x cm)	1.9 m^2	2.25 m^2	0.8 m^2	9.6 m^2
Concentr.factor	none	3.5	weak	1.4
Intensify.mirror	none	8 (funnel-shaped)	behind the pipes	1
Weight (kg)	120 kg	120 + 130 kg	75 kg	no info.
Portability	stationary	disassemblable	rolls or slides	stationary
Heating power (1 l H_2O, 20-100°C)	10 liter: 20-100°C for store temp. = 140-100°C	on hot stove: 10 sec.	up to 1 kW	186 liters up to 80-100°C in 65 min.; 950 W/m^2
Efficiency (partial effic.)	collector + store: 30 %, 90 K	store heating: 52 % 150°C 850 W/m^2	tubular collector: 54%, 100 K, 850 W/m^2	43 %
Max. temperature	165°C (25°C amb.)	300°C (20°C amb.)	220°C (25°C amb.)	over 100°C
Qty. of food (kg)	5 kg, later 10 kg	7 kg	no info.	approx. 200
No. of vessels	1, later 2	1	3, only 1 hot plate	7
Tea/baked goods	yes/yes	yes/yes	yes/some	yes/no
Roasting/grilling	no/no	yes/no	yes/yes	no/no
Keeping warm	approx. 12 h	approx. 12 h	yes, overnight	more than 1 night
Must cook stand in sun?	no	no	no	no
Stirring possible?	yes	yes	yes	yes
Stirring neces.?	sometimes	yes	sometimes	no
Solar tracking?	no	every 60 min.	no	no
Flare?	no	no	no	no
Blinding?	no	no	no	hardly
Burns?	hardly	yes	no	no
Special disadvantages	immobile, elaborate	handling of hot slabs, 2 sep. units	sensitive tubes, limit. storage cap.	no morning/evening use, stationary pots
Special advantages	stationary, evening cooking and in the shade	indoor cooking and in the evening	adjust. heat output (e.g. for flat cakes), dome	air as heat conductor
General handling	very good	complicated	good, easy	easy, autom.
Manufacture	by craftsmen	industrial	industrial/ handicraft	by craftsmen
Requisite handicraft skills	metalworking, accurate plumbing	precision mechanics, sheet metalworking	various metalworking skills	metalworking
Requisite tools and fixtures	brazing unit, metalworking and plumbing tools	folder/bender, welding unit, large tin snips	machine-shop tools	machine-shop tools
Materials not locally available	foil, absorber coating, Cu-tubes, flat pot	fluoropolymer foil, alum. reflector, mineral fibers	collector and special metal parts, flat pot	safety glass, absorber foil, alum. reflector foil
Weak points	glass, reflector	reflector, foil	collector pipes	none
Remedy by	replace glass/ foil by hand	replace by hand	replace with spare pipe	no info.
Service life	approx. 10 years	approx. 10 years	approx. 10 years	target: 10 years
Cost of manuf. US$	approx. $ 1.400 (in small lots)	$ 400 (in large lots)	buy: $ 1.700 (10 ea.)	no info.
Purchase price US$	no info.	no info.	$ 450 (10.000ea.)	no info.
Cost of maint. US$/a	$ 20	$ 40	$ 40	no info.
PEL (USc/kWh)	28 c	20 c	48 c	-
CC, CCL (USc/kg)	5.9 c (CCL)	6.3 c (CCL)	15.6 c (CCL)	-
Target group (acc. to manuf.)	Schools, small inst., hospitals	all groups in devel. countries	upper class; all in devel. count.	urban + rural institutions

Appendix 2: Self-construction Directions for the ULOG Tropical Solar Cooker

ULOG SOLAR-COOKER/OVEN

TROPICAL MODEL 85

SELF-CONSTRUCTION DIRECTIONS

This instruction booklet is available direct from the author (see below). Price in Switzerland: Sfr. 16.--, payable to Swiss PTT postal giro account BASLE 40-18151-8; if ordered from FR Germany: DM 20.--, by registered letter. Prices include postage of booklet.

Ulrich Oehler, Development Engineer
Morgartenring 18, CH-4054 Basle, Switzerland
Tel. 061 - 38 66 22

(p.3)

ULOG - Solar-Cooker/Oven - Tropical Model 85

Parts List

Item	No.	Description	Dimensions	Material
01	2	Sheet of glass	50 x 50 x 0.2	Window glass
02	1	Stove bottom (trough)	79 x 79 x 0.03	Aluminium sheet/offset plate
03	4	Frame for stove bottom	66.5 x 8 x 3	Wood
04	4	Frame for glass	55.5 x 6 x 2.5	Wood
05	4	Glass-frame spacer battens	49.2 x 2.6 x 1.3	Wood
06	8	Glass-frame retaining strips	49.2 x 1.3 x 1.3	Wood
07	1	Reflector foil	56 x 56 x 0.01	Silver paper, tinfoil
08		Adhesive		for item 7
09	1	Floor	67.5 x 67.5 x 0.5	Plywood
10	4	Side wall	67 x 25 x 0.5	Plywood
11	2	Reinforcement	55 x 4.5 x 0.5	Plywood
12	1	Cover (lid)	56 x 56 x 0.8	Plywood
13	4	Hinge	10 x 5 x 0.2	Leather or textile belt(ing)
14	4	Leg	35 x 4 x 4	Wood
15	4	Reinforcing strips	58.4 x 2.5 x 1.5	Wood
16	4	Revetment (lining)	61,5 x 6 x 1.5	Wood
17	1	Cord tightener	5 x 2 x 0.5	Hardwood
18	1	Cord	Ø 0.2 x 120	Hemp or jute
19	1	Prop for lid	Ø 2 x 56	Wood, tree branch
20	3	Handle	Ø 3 x 36	Wood, tree branch
21	18	Nail	Ø 0.2 x 5	Iron / steel
22	108	Nail	Ø 0.15 x 3.5	Iron / steel
23	98	Nail	Ø 0.1 x 2	Iron / steel
24	12	Wood screw	Ø 0.35 x 5	Iron / steel
25	-	Insulation material	coconut fibre (coir)	Ground-nut (peanut) husks Dry hay
26	-	Paint/colouring (black)	Heatproof Non-poisonous	or charcoal, soot, carbon black

(p.4)

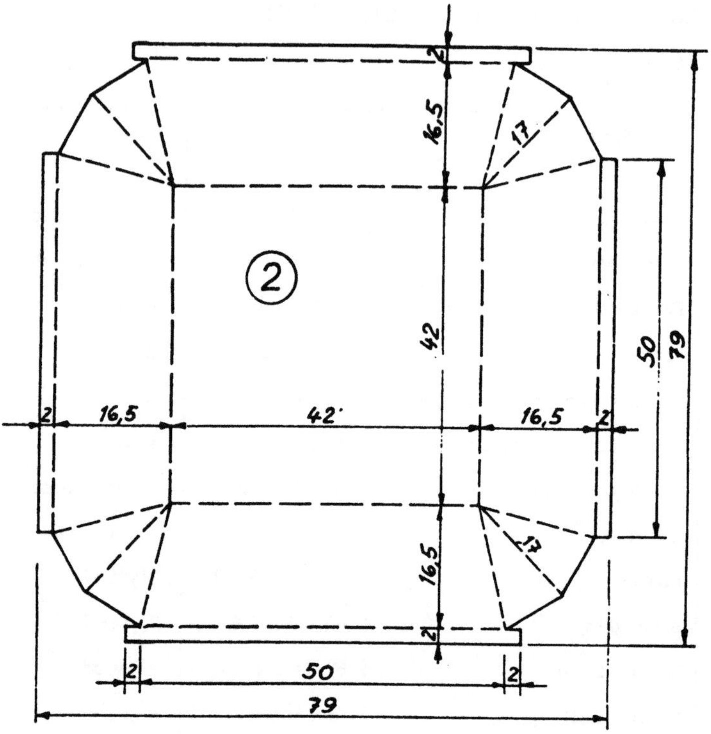
2
2
16,5
17
42
50
79
2
16,5
42
16,5
2
17
16,5
2
2
50
2
79

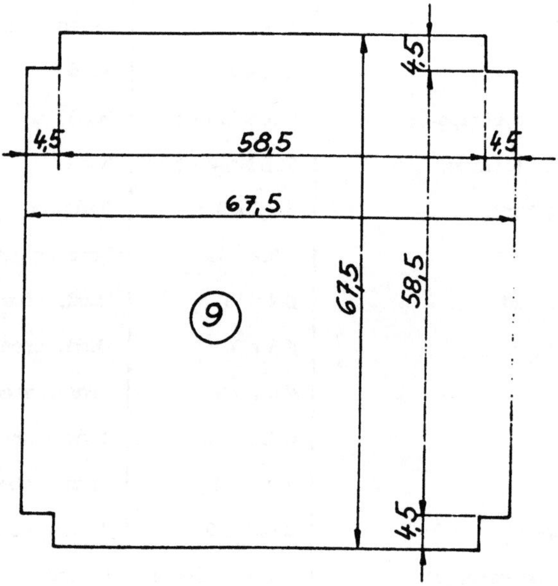
4,5
4,5
58,5
4,5
67,5
67,5
58,5
9
4,5

(p.5)

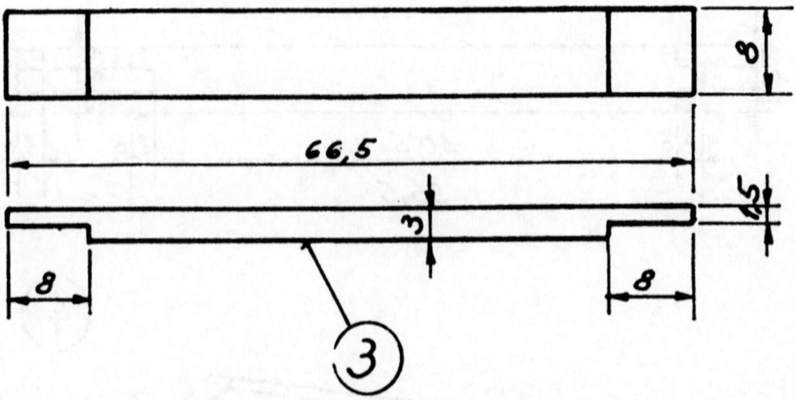

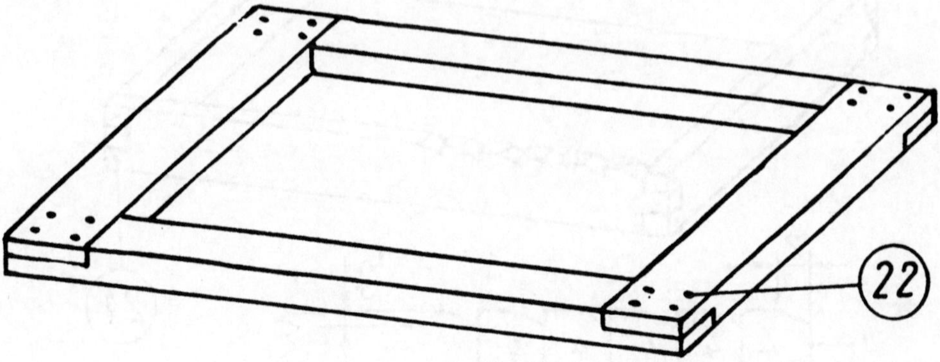

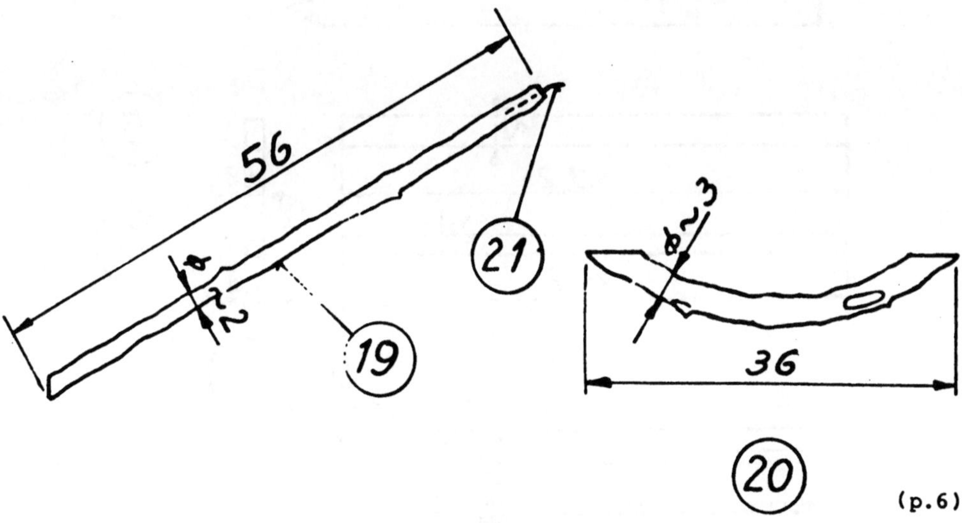

(p.6)

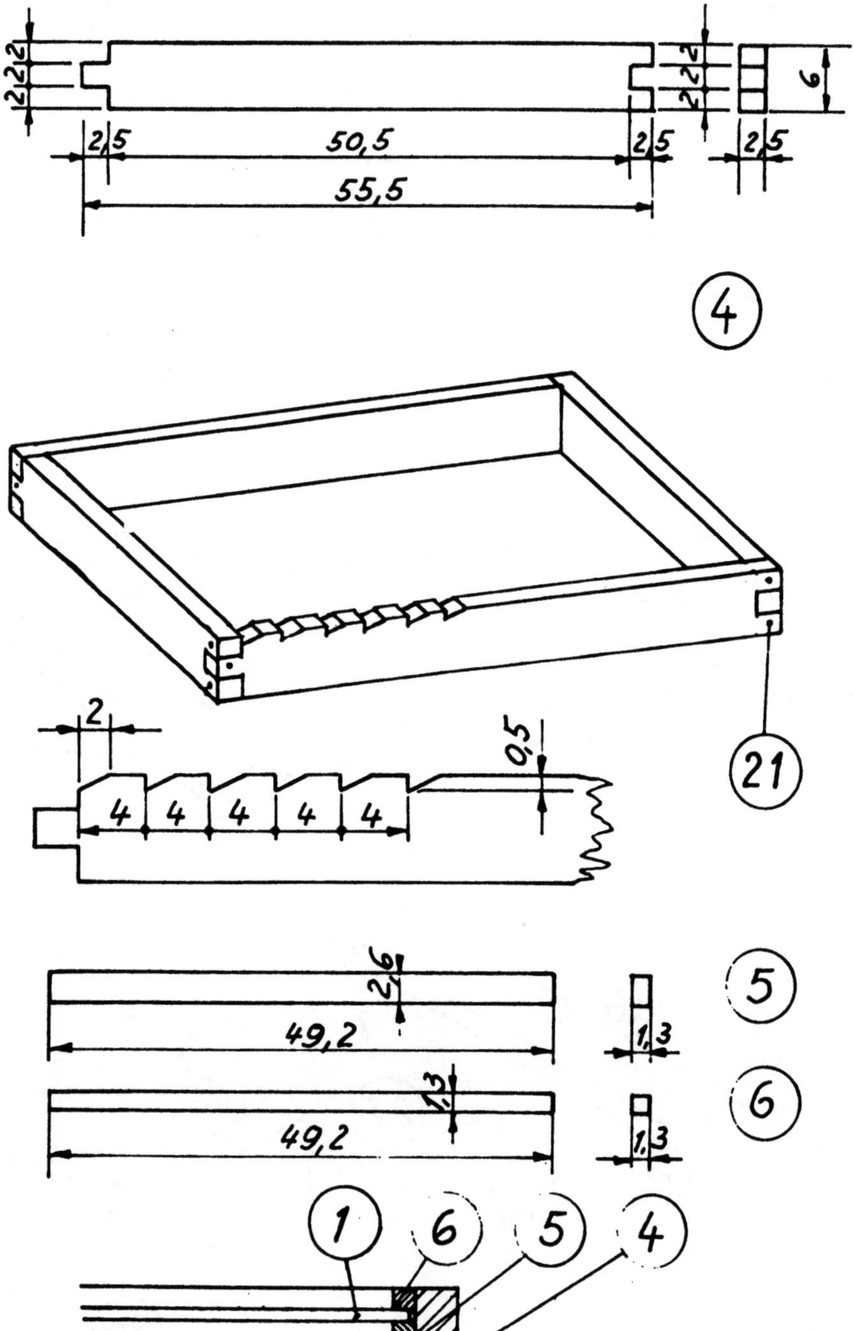
2
2
2
2
2
2
6
2,5
50,5
2,5
2,5
55,5
4
21
2
0,5
4
4
4
4
4
5
2,6
49,2
1,3
6
1,3
49,2
1,3
1
6
5
4

(p.7)

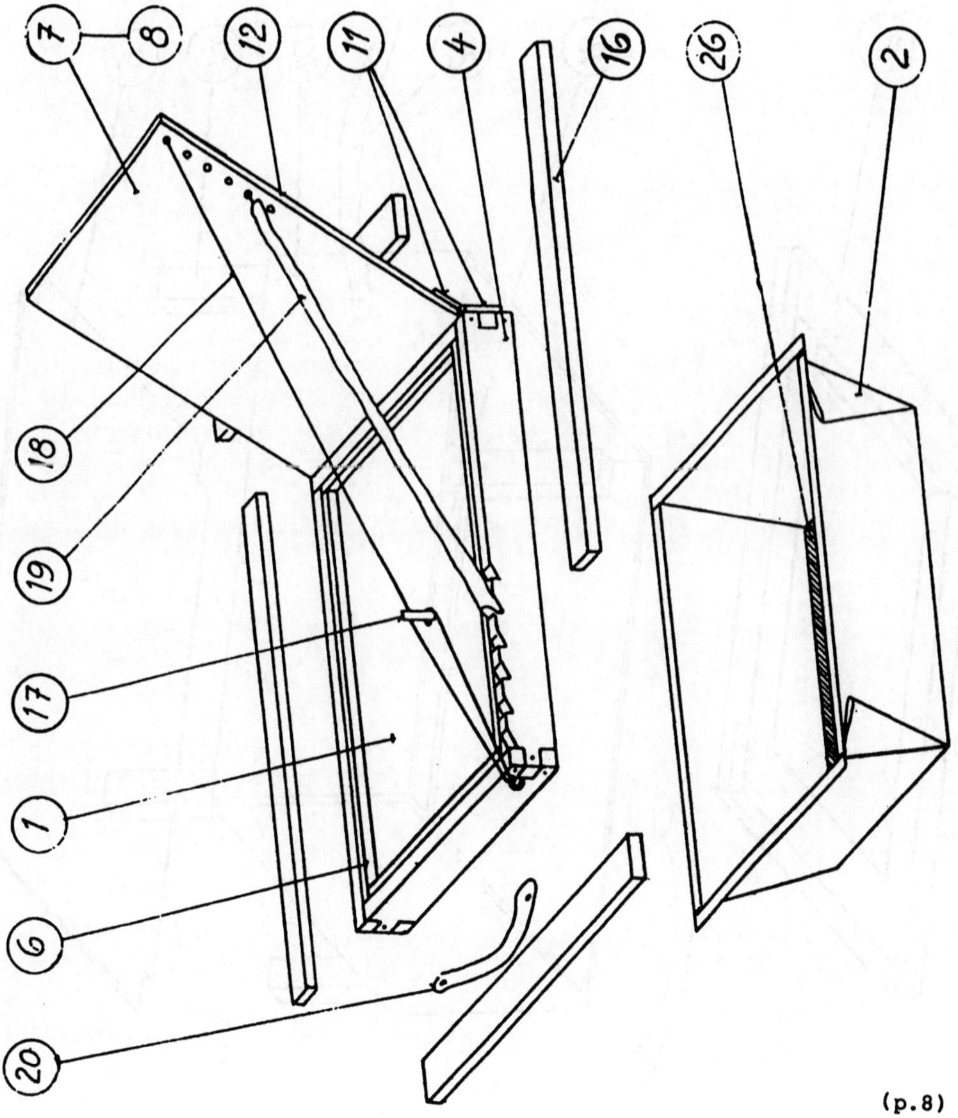

(p.8)

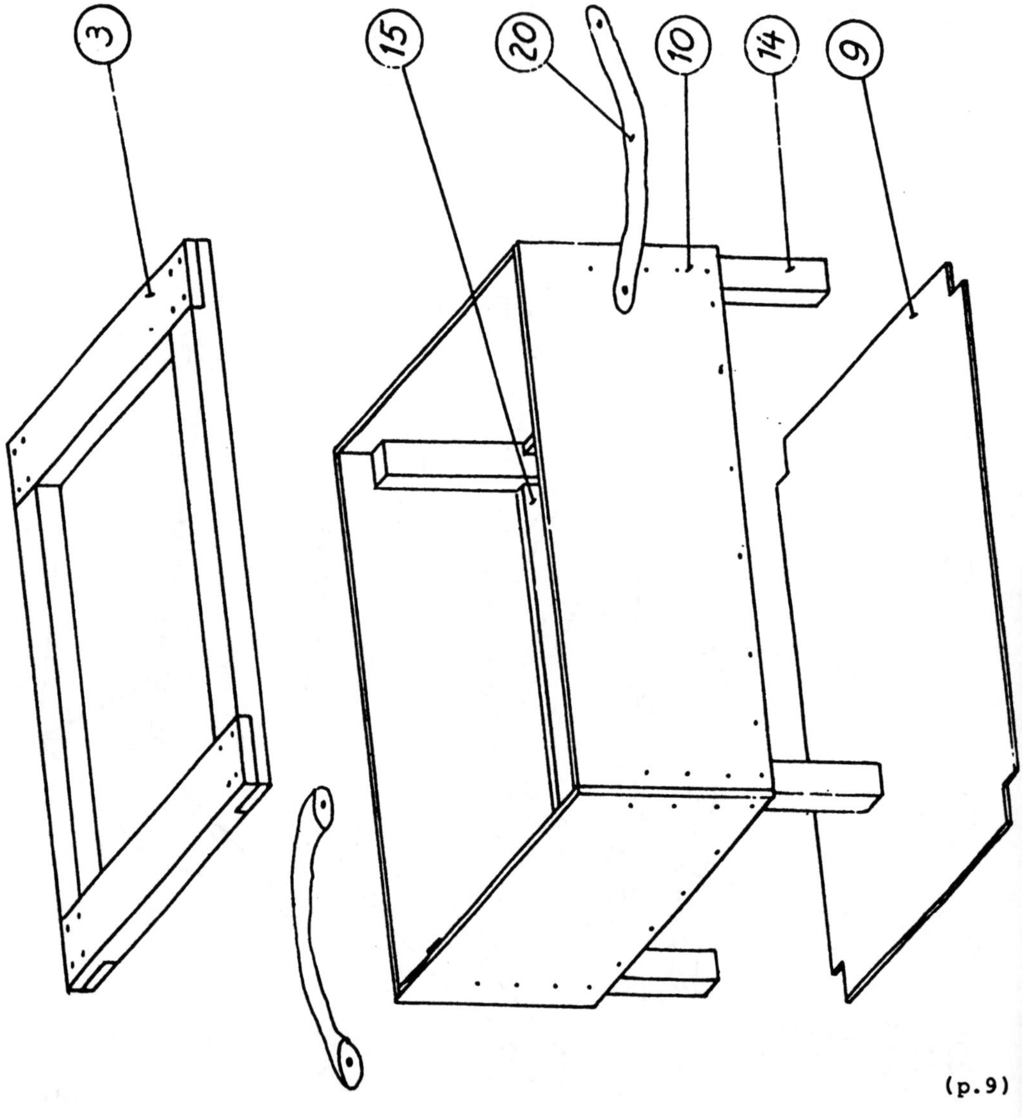

(p.9)

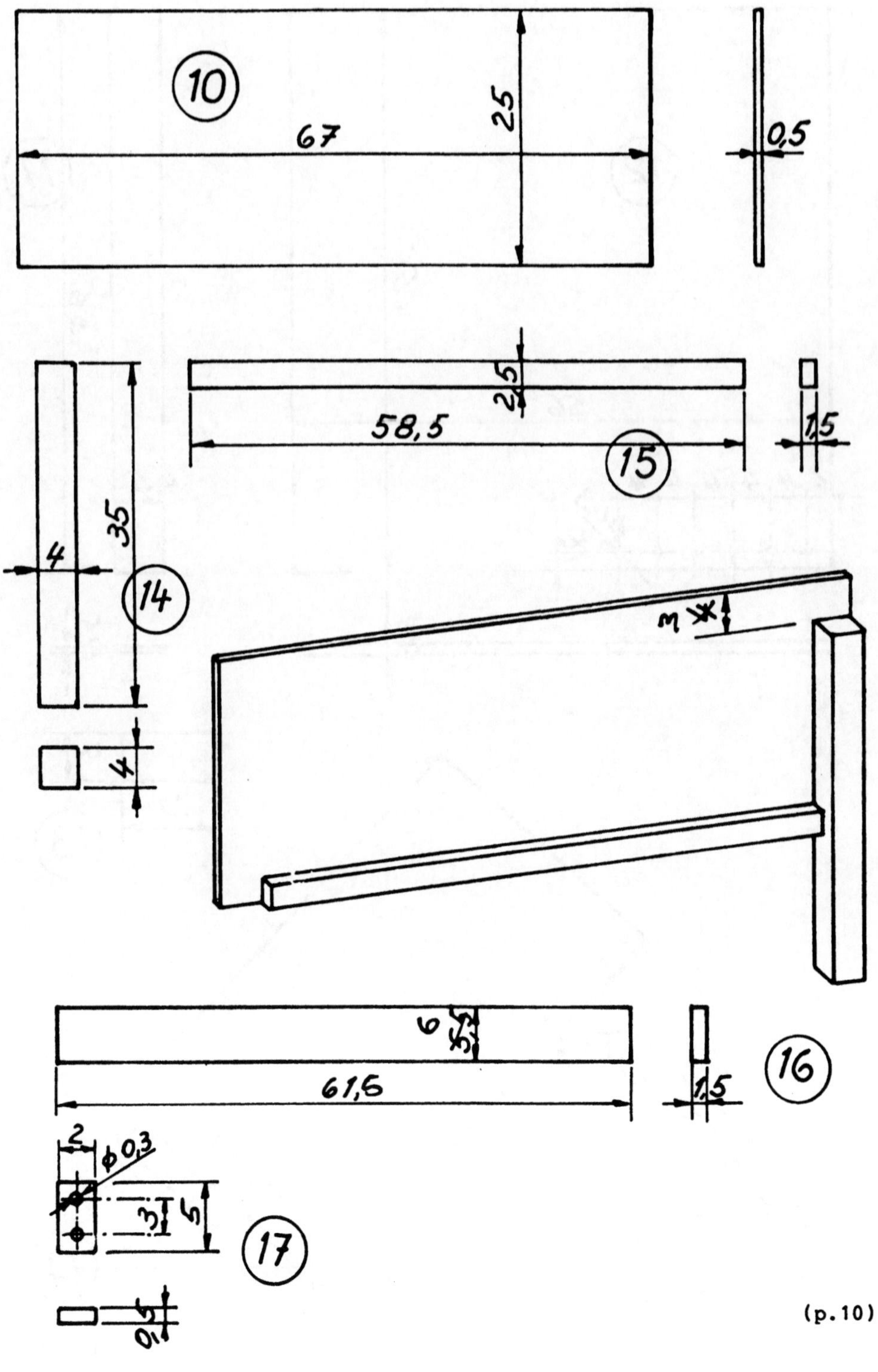

(p.10)

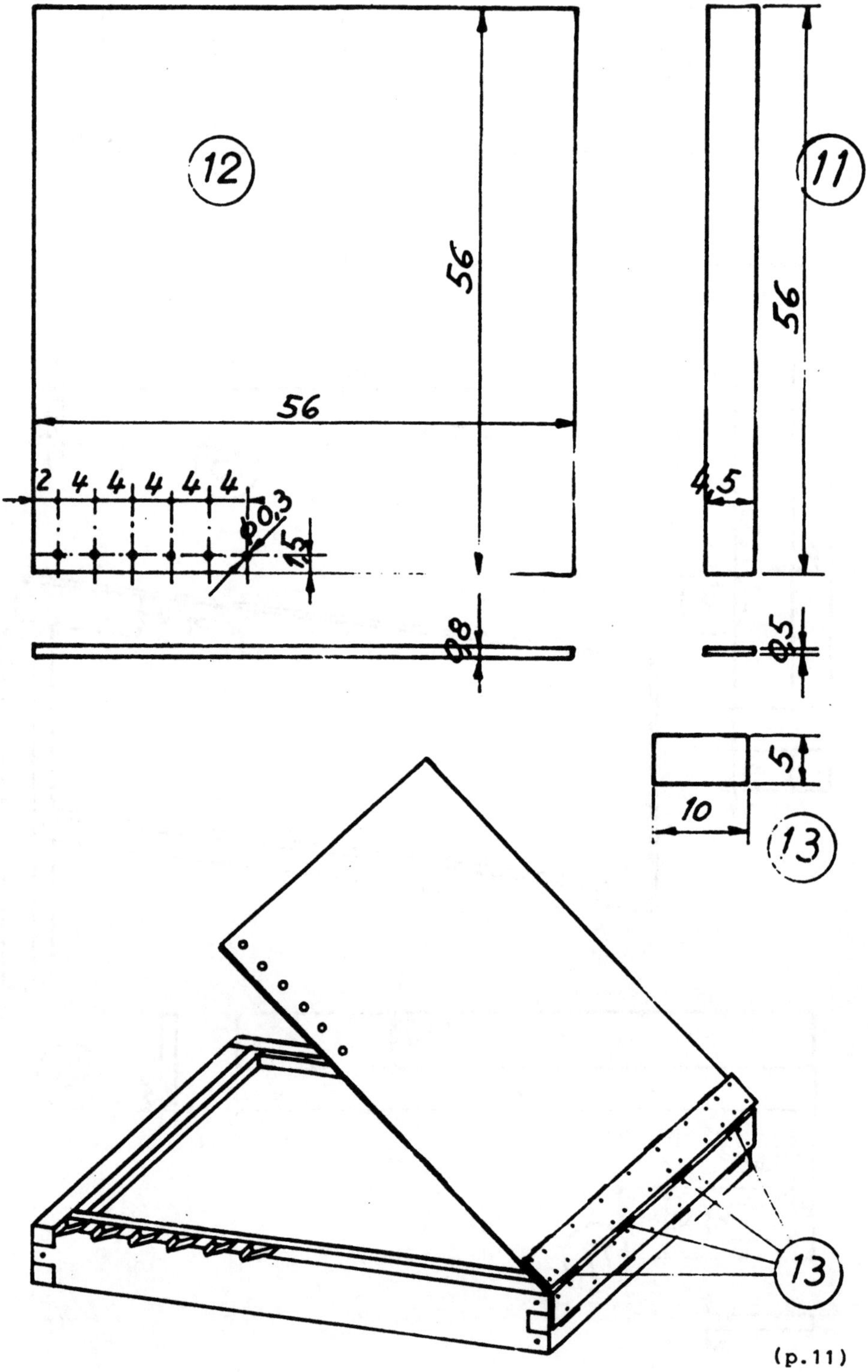

(p.11)

Construction Directions

Collect the materials according to the Parts List (p. 4) and prepare all 26 items in sufficient quantities and/or in the dimensions given.

If window-glass panes of 2mm thickness cannot be procured, thicker panes (3-6mm), e.g. from old cars, can be used. Particularly where children playing could cause glass breakage it is advisable to use breakproof material (hardened or toughened glass, laminated or compound glass, fluorine-containing plastics sheet) for the upper sheet of glass. The glass-frame spacer battens (5) may have to be planed down thinner if thicker (3-6mm) panes are used.

All four battens for the glass-frame (4) should have one tenon at one end and two tenons at the other. These can be cut with saw, firmer chisel and wood rasp. The four battens are joined, each by a nail (21) in each of the 12 tenons, to the glass frame which should be right-angled and distortion-free and which should show no projecting cogs. Now, or even already before the four battens are nailed together, one of them should be provided with six notches according to the dimensions shown on p.7.

Then four glass-frame retaining strips (6) are fixed, each with 2 or 3 nails (22), exactly parallel to the frame, without projecting cogs; one sheet of glass (1) is carefully cleaned and inserted; four spacer battens (5) are also fixed, each with two nails (22); the second glass pane is cleaned and inserted; and, finally, the remaining four retaining strips are put into place and fixed.

On the lid (12), the edges should be slightly rounded down with a file or sandpaper, the four hinges (13) then being laid in place, spaced at regular intervals apart, with reinforcement in the shape of a fastening plate (11), the whole being nailed together as shown on p. 11. For this you use 22 nails (23) which are turned in and nailed into the plywood surface. Then glue the reflector foil over the whole inside surface (underside) of the lid and drill the six holes for the lid prop. The finished lid should then be fitted, with the second reinforcement (11) and 20 nails (23), to the upper edge of the glass frame so that it closes firmly by its own weight with as little play as possible (see drawing p. 11).

The frame (3) for the stove bottom is cut to size according to p. 6 and nailed together at right angles, free of distortion and without projecting cogs, with 12 nails (22). Nails protruding on the underside are hammered down flat.

In accordance with p. 10, one leg (14) should be fixed to each side wall (10) with 3 or 4 nails (22) as well as a reinforcing strip (15), fixed with five nails (23). Then the four side walls are nailed - each with four nails (22) - to the frame (3) and, each with 3 or 4 nails (22), to the corresponding legs. To strengthen the structure, hammer a nail (21) perpendicularly through the frame (3) into each of the four legs (14).

Cut the stove bottom (trough) (2) exactly to size according to the drawing on p. 5 and lay it on a table with the shiny side upwards. Press a sharp-edged, straight piece of wood (e.g. a plane) of a maximum of 42cm in length on the square drawn on the middle of the sheet so that one side of the stove bottom can be bent upwards exactly along the broken line. When all sides have been thus bent upwards you can form a corner by pressing the two points one against the other, which should shape a corner of the upper frame of the stove bottom. Now, but not before, you fold the three-cornered "peak" along the 17cm-long broken line sketched in on the drawing. Then, turn or beat down the folded "peak" and press it on to one side wall of the stove bottom (trough) as shown on p. 8. After all four corners have been shaped in this way, you then bend the four upper edges of the sheet over at right-angles (over a straight tabletop edge, for example). Lay the finished stove bottom in the frame (3) and fix it with 16 nails (23).

(p.12)

After the trough has been fitted, place the whole cooker upside down on the floor so that the four legs are uppermost. Completely fill up all spaces, hollows and cavities with insulation material (25) which must be absolutely dry; it should not be compressed but must be filled so tightly as to ensure that no hollow spaces remain and that none can arise when, after the floor (9) of the cooker has been fitted, the cooker is set firmly on its legs. To prevent the stove bottom from sagging under the weight of a full saucepan, it is advisable to insulate the middle of the stove with compression-resistant materials such as, for example, several layers of corrugated board, egg cartons and the like. Now the cooker, with the floor (9), is closed by means of 20 nails (23) and then placed on its legs.

To place the glass frame (4) centrally over the stove bottom (2), nail the revetment or liner casing (16) on the frame (3) with 16 nails (22). This will improve thermal insulation.

Now the shiny 42 x 42cm floor of the trough should be coated with matt, black, heat-resistant paint (26). Since most ready-to-use paints contain toxic solvents, this should be done outdoors or near an open window. For the same reason the cooker should not be used for cooking food until several days after completion and not until it has first been heated up many times. Paint which is too fresh can adversely affect the taste and smell of the food. The floor of the stove bottom can also be covered with small pieces of charcoal or sprinkled all over (no gaps) with charcoal powder, instead of a coat of paint.

A suitably-shaped branch is screwed on to the front of the glass frame (4) with 2-4 wood screws (24), to form a handle (20) and to enable the frame to be easily lifted up for a saucepan to be put in the cooker. Make quite sure when fixing the handle that the points of the screws do not touch the glass. The other two handles (20) are likewise screwed on to the sides of the cooker, each with 2-4 wood screws (24) so that, in this case, the screws pass through into the frame of the trough (3). Two persons can easily carry the solar cooker/oven by the side handles.

The lid prop (19) also consists of a tree branch. A nail (21) is hammered into the centre of one end of the prop but only so far as to leave about 15mm still protruding. With a file, remove the head of the nail and round off the stump with a file. The protruding nail should then be slightly bent over so that it fits easily into any one of the holes in the lid.

Remove the head of a nail (21) and file the stump to a point. Then shape the nail into a "U". The resulting pointed staple is hammered into the front, right-hand end of the glass frame and serves as an eyelet for the cord (18) with cord tightener (17) as shown on p. 8 and the front page.

(p.13)

Items 9, 10, 14 and 15 can be of metal instead of wood and/or plywood, if metal is easier to come by; for example, a sheet-metal drum cut in half lengthwise can serve as a trough for the stove-bottom frame. However, strict attention should be paid to the space between the metal trough and the stove bottom which, everywhere, should be at least 8cm. It is also imperative to ensure that the space and any hollows be completely filled with insulating material. Two or three cookers can be built in to a correspondingly long trough, side by side, in order to cook larger amounts of food as well as to reduce heat loss.

General Notes

This solar oven is primarily suitable for cooking and baking in those countries lying between the two tropics, i.e. near the equator. The designer has developed and tested other models for other countries.

At midday, the reflector is best positioned at about 20°-30° to the rear, as shown on the front page. Two hours before or after midday it should be more or less perpendicular, whereas in the morning and evening a leaning-forward position is better. In this way the foil reflects the sunrays falling on it into the cooker where they intensify the direct, incident sunlight. Further reflectors all round the glass frame can enhance this effect. Their most effective angular setting at any given time can easily be determined by trial.

All dark-coloured pots, pans and cake tins are suitable as cooking and baking utensils. Most important is that lids should be matt black, as far as possible. Since hot air is lighter than cold, it is hotter in the upper half of the stove bottom (trough) than in the lower. It is therefore advisable to use cooking vessels which are as flat as possible and to place them on a black-painted, wire or wooden grid or on a layer of charcoal so that they are close to the glass-sheet. If the hot air can also circulate underneath the cooking vessel the food is heated up quicker and more evenly.

Large amounts of food, for families with many children, school kitchens, etc., are prepared in larger solar cookers but, whatever size is used, the depth of the stove bottom (trough) should not be greater than that of the ULOG 85 model described herein. Since large glass panes are difficult to come by and to transport, and their replacement in case of breakage is quite expensive, it is recommended that several standard cookers be used instead of an extra-large one. If you build two or three cookers together in a row, heat loss (and thus cooking time) is reduced. On the other hand, a large slow cooker (similar to a low-heat hay box cooker) can be useful for preparing large quantities of food. It consists of a large, if possible ball-shaped (length, depth and width the same) vessel which can stand in a larger crate or tub allowing for at least 10cm of space (between cooking vessel and crate) on all sides. This interspace is filled up all round and under the vessel with insulation material. As soon as the sun shines, cooking is started in all already existing and available solar cookers. Then, as soon as the food has reached boiling-point it is filled into the slow cooker which must be immediately but carefully closed with a tight-fitting insulating lid. Now, the second and third helpings can be heated or boiled up in the solar cookers, and the hot food again goes into the slow cooker. Like this, a swarm of hungry people can be well fed in the evening.

After sundown it is still possible to have hot meals from the solar cooker if, at dusk, the lid (12) is closed and the cooker is covered up with a blanket. The food will then keep warm for a few hours longer but it can be kept warm longer still in the slow cooker described above.

Cooking time can be determined by means of the following data:
When the sun is at the zenith and its radiation is at right-angles, it emits energy amounting to 1 kW on one square metre, given a cloudless sky and clear air. Fog, mist, haze, dust, wind-blown sand and clouds have considerable influence on the effective quantity of energy.

The solar cooker/oven described in this booklet has an effective glass surface area of 0.23 m^2. If clean and set at the correct angle, the reflector foil can reflect about 25% of the light into the cooker. Thus, under optimum conditions, the user can reckon on a total of 288 watts; 1.16 watt-hours are needed to heat up one litre of water by 1°C. Heating-up from an ambient temperature of 25°C to 95°C boiling temperature means a temperature rise of 70 degrees. So, to boil up a litre of water you need 70 x 1.16 = 81 Wh, which in our solar cooker takes 81 ÷ 288 = 0.28 hours, or 17 minutes. However, since not only the water but also the vessel and the interior are being heated, and since some loss of reflection, and thus radiation, occurs, the water will boil only after 25-30 minutes. About 50-60 minutes are needed to boil two litres of water.

Concentrating, or focusing, solar cookers offer another possibility of increasing the volume of food to be cooked. They work with a concave parabolic reflector mirror and can heat a large cooking pot from below (as with a conventional hot plate) and the depth of the pot is not limited to 16cm. But this solar-cooking principle is not without its disadvantages and hazards. The typical features of the two systems are compared in the juxtaposition which follows.

Recipes for cooking in the solar oven are very numerous: in the USA alone at least three solar-cookery books have been published. But it is fun to experiment and we would not wish to deny "solar" cooks this pleasure. We therefore give only a few guidelines, as follows:

The solar cooker should never be opened during the cooking or baking process - otherwise heat gets lost. But this also means that only so much moisture escapes from the food to be cooked or baked as to saturate the air space in the cooker with water vapour; afterwards, the food hardly dries out any more. Generally, therefore, less water is needed than with conventional cooking. For example, if you have been used to cooking two parts of rice with three parts of water, you should now choose a ratio of 1:1. Thus, you can cook tea, soups or sauces with advantage in very concentrated form and, before serving, dilute them with hot water. By this means you can either shorten cooking time or increase the amount of food to be cooked. If cereals, pulses, legumes, dried vegetables or desiccated fruits are put in clean water the day before, they will "slow" cook much quicker in the ULOG solar cooker/oven.

MUCH FUN AND BON APPETIT !

Please address all enquiries about design plans, prices for finished products and dates of courses in solar-cooker construction to:

Ulrich Oehler, Development Engineer,
Morgartenring 18, 4054 Basle, Switzerland
Tel. 061 - 38 66 22

(p.15)

SUPPLEMENTS FOR THE DOUBLE MODEL

For certain users the described solar cooker can be too small. In this case, we recommend to employ, instead of two cookers, a double model. In what follows are described the changements in relation to the simple model. The dimensions in one direction are all increased of **53 cm.**

Changements in the parts list **(bold)**

Item	No.	Description	Dimensions
1	**4**	sheet of glass	
2	1	stove bottom (trough)	**132** x **79** x 0.03
3A	2	frame for stove bottom	**119.5** x 8 x 3
3B	2	"	**66.5** x 8 x 3
4A	2	frame for glass	**108.5** x 6 x 2.5
4B	**3**	"	55.5 x 6 x 2.5
5	**8**	glass-frame spacer battens	
6	**16**	glass-frame retaining strips	
7	1	reflector foil	**109** x 56 x 0.01
9	1	floor	**120.5** x 67.5 x 0.5
10A	2	side wall	**120** x 25 x 0.5
10B	2	"	67 x 25 x 0.5
11	2	reinforcement	**108** x 4.5 x 0.5
12	1	cover (lid)	**109** x 56 x 0.8
13	**8**	hinge	
15A	2	reinforcing strip	**111.4** x 2.5 x 1.5
15B	2	"	58.4 x 2.5 x 1.5
16A	2	revetment (lining)	**114.5** x 6 x 1.5
16B	2	"	61.5 x 6 x 1.5
17	2	cord tightener	
18	2	cord	
19	2	prop for lid	

If offset plates of the good size can't be found for the trough, it's possible to assemble several plates only by folding as shown below (without soldering or gluing). This kind of joint is complet- ly thick to vapor.

For the glass-cover we don't use sheets of glass of the double lenght but the same size as for the simple model. Two of them are put side by side separated by a piece of frame (item 4B). To join this with the long pieces of frame (item 4A) a tenon has to be cut on both ends in the middle and a suitable hole in both long pieces of frame (see figure below).

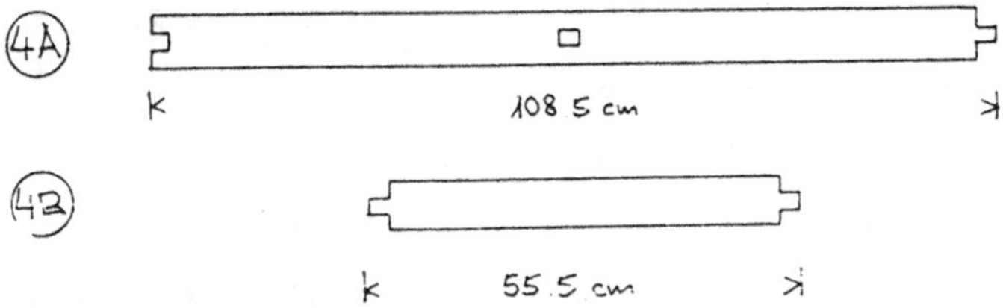

It'a a advantage to hold the reflector with two props. In the same time, also two cords and tighteners are needed. All the rest can be made analogously to the instructions of the simple model.

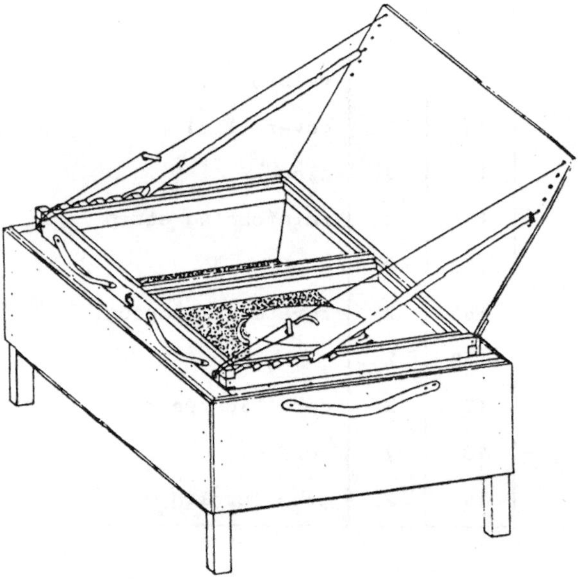

GPSR Compliance

The European Union's (EU) General Product Safety Regulation (GPSR) is a set of rules that requires consumer products to be safe and our obligations to ensure this.

If you have any concerns about our products, you can contact us on ProductSafety@springernature.com

In case Publisher is established outside the EU, the EU authorized representative is:

Springer Nature Customer Service Center GmbH
Europaplatz 3
69115 Heidelberg, Germany

Batch number: 10815974

Printed by Printforce, the Netherlands